Hertfordshire
COUNTY COUNCIL
Community Information

Please renew/return this item by the last date shown.

So that your telephone call is charged at local rate,
please call the numbers as set out below:

	From Area codes 01923 or 0208:	From the rest of Herts:
Renewals:	01923 471373	01438 737373
Enquiries:	01923 471333	01438 737333
Minicom:	01923 471599	01438 737599

OVS
796.
525
BED

L32b

UNDERGROUND BRITAIN

UNDERGROUND BRITAIN

A Guide to the Wild Caves and Show Caves of England, Scotland and Wales

BRUCE BEDFORD

WILLOW BOOKS
COLLINS
8 Grafton Street, London
1985

Willow Books
William Collins Sons & Co Ltd
London · Glasgow · Sydney · Auckland
Toronto · Johannesburg

First published in Great Britain 1985

Bedford, Bruce
Underground Britain
1. Speleology
I. Title

796.5′25′0941 GB608.43

ISBN 0 00 218101 0

Designed and produced by
Robert Adkinson Limited, London
Editorial Director   Clare Howell
Editor   Christopher Pick
Art Director   Christine Simmonds
Designer   Mike Rose
Illustrations   Rick Blakely
Cartography   Kirby-Wood

Phototypeset by Ashwell Print Services,
Ashwellthorpe

Illustrations originated by La Cromolito,
Milan

Printed and bound by Graficromo, Cordoba,
Spain

CONTENTS

Introduction

Our attitude to underground places today is markedly different from that of our earliest ancestors. Whereas caves were once regarded as our natural and safest habitat, we now tend to look upon them with awe, and only insatiable curiosity drives us to descend underground to see what they are really like.

Perhaps it is all part of the desire for out-of-the-ordinary experiences: the desire, shared by tourist and sportsman alike, to get off the beaten track and away from the madding crowd. It is to satisfy this innate curiosity, at least in part, that this book was written. I say in part, because the book should also excite that curiosity even more. Then, I hope, there will be only one choice for the reader previously unacquainted with underground Britain: to go out and visit this strange, awesome and often exotically beautiful kingdom of the dark.

We are fortunate in Britain in possessing a considerable number of both fine show caves and challenging wild caves – those aptly named domains of the experienced caver or potholer.

Show caves give the public the chance to walk through the marvellously varied passages and great chambers of the subterranean world, and to see at first hand the exquisite formations with which nature has decorated them. These are sights usually enjoyed only by cavers, but in show caves the dangers and obstacles have been removed or by-passed, and, with floodlights picking out every detail, the visitor can stroll at leisure along easy pathways in complete safety.

The descriptions of some of the major and most interesting wild caves throughout the country are – it must be said – for the armchair explorer. But it is my hope that they will give some indication of the stimulation and challenge – not just physical but mental as well – that caving provides. British cavers do not, as a rule, actively seek new adherents. They much prefer that newcomers should approach the sport from a spontaneous and self-generated interest. Those who do will find a welcome in the caving fraternity.

Caves apart, we have exploited the underworld in many different ways over thousands of years, working chambers and passages out of the rock beneath our feet out of fancy or sheer necessity. From the humblest cellars to the most spectacular slate mines, such places tend to be far more permanent than those structures we erect above ground – and they all have for us the fascination of the secret, the hidden, the subterranean. Some have private access only, but there are many to which the public is not only admitted but positively welcomed. Major man-made sites open to the public are also described in detail, and many minor sites too.

This book offers the key to the hidden world of underground Britain. I hope it unlocks many fascinating, secret places for you.

BRUCE BEDFORD

What is a cave?

What is a cave? For those who already know, of course, the answer seems self-evident. But they are exceptional places in the sense that the great majority of people may only encounter one in their lifetime, if even one — and that one is more than likely to be a show cave, with its attendant concrete paths, bright electric illumination, handrails, bridges, and all manner of other aids to easy access.

Strictly speaking, the word 'cave' should be applied only to natural cavities in rock, and it is generally accepted that they should be large enough to allow the passage of a human being. At a number of sites described in this book, the word 'cave' is used to denote almost any sizable man-made excavation, although it would be more correct to refer to these by some other name, such as tunnel or mine. But to be too pedantic on this point would only result in confusion.

True caves are natural phenomena. In volcanically active countries such as Hawaii and Iceland, there are lava caves, formed when long tongues of lava cool on the outside but continue to flow on the inside, leaving hollow rock tubes sometimes miles long. Elsewhere there are ice caves too, formed by summer meltwater streams gushing through the depths of glaciers, leaving systems with their own unique challenges and hazards for alpine cavers.

In Britain there are three other types: fracture or tectonic caves, sea caves, and inland caves. The first embraces fissures in rock formed when blocks of rock moved relative to one another, leaving a gap between them. Such fissures are usually both small and short, and are of little interest to cavers.

Sea caves are formed when wave and tidal action erodes weak areas in the rock until passages and chambers are formed. In Scotland and the south-west coastline of England, for instance, the result can be quite dramatic entrances piercing the tall cliffs, although their length is limited by the topmost reaches of the tides. Where the receding sea has left an old coastline high and dry, such sea caves can also be found some distance inland, as on Arran.

However, it is inland caves that constitute the major category and give such a wonderful variety of shapes and sizes, throughout the world, from Jean Bernard in France, more than 5100 feet deep, to the 294-mile-long Mammoth Cave in the USA. For comparison, the British records are held by Ogof Ffynnon Ddu, in Wales, which is just over 1000 feet deep, and the Lancaster-Easegill complex, at the junction of Lancashire, Cumbria and North Yorkshire, where there are some 30 miles of passages.

Left The Bridge, Dowbergill Passage, Yorkshire.

Right The Entrance to Ogof Rhyd Sych.

How inland caves are formed

Inland caves depend for their existence on three things: soluble rock, rainwater and time. The most suitable rock is limestone, which is soluble in slightly acidic water and is also plentiful, constituting about a twelfth of the world's total land area. Its solubility stems from its composition, for it is made up in the main of the skeletons of countless billions of tiny sea creatures. These skeletons accumulated on the floors of the great oceans as long as 400 million years ago, though the process continues today in the warmer seas. The accumulating weight of these vast sub-aquatic graveyards compressed the lower layers into solid beds of limestone, many of which were later raised by the restless movements of the planet's surface until they stood above sea level.

Thus lifted and exposed, the beds of limestone would have produced little of interest for us — except pleasant hills and mountains — were it not for the fact that in the process the beds became fractured by horizontal and vertical cracks. Into these cracks seeps rainwater, which by taking up a tiny amount of carbon dioxide from the atmosphere becomes very slightly acidic. Where the rainwater has to seep through soil before reaching the limestone its acidity is increased considerably.

Given access to the heart of the limestone, the water begins its twin work of destruction and construction. Slowly the cracks are enlarged by corrosion until the critical stage is reached at which the slightly larger fissures are extensive enough to begin to take the main flow of water. When this happens, their growth accelerates until they are taking nearly all the water from neighbouring cracks. Eventually they will reach cave size — but of course they are completely flooded.

Fortunately for us, water levels change over the years. A valley in which the cave water springs up to the surface will be worn ever deeper, eventually draining the higher flooded section of limestone. The caves formed in the higher area are left partially or completely dry, depending on whether the cave streams are still flowing along the floors of the passages or have found some lower course. Cave passages abandoned by the streams which formed them are known as fossil or dead caves; active caves are those in which the stream still flows.

The first stage of development, when the entire passage is flooded (phreatic

Above Town Drain cave in flood conditions.

Right Waterfall, Piccadilly — Ogof Ffynnon Ddu II.

development), produces a distinctive round or elliptical cross-section of passage. The second stage (vadose development), when the stream runs along the floor leaving an air-space above, results in equally distinctive sharp down-cutting, which is why so many passages have a keyhole cross-section.

Vertical shafts occur most commonly where the limestone beds are more or less horizontal, less often where the rock strata dip sharply — as on Mendip, where the tilt is about 45 degrees. As the water enlarges ever lower passages, a stream will sometimes abandon its slow horizontal flow to open up a vertical weakness in the rock, resulting in a shaft (or pothole).

Chambers are formed when particularly weak areas in the rock collapse into the streamway below, the debris being reduced by the water and carried away. When a chamber becomes dead, after the stream has taken a lower course, its floor can be left littered with often massive boulders.

These, then, are the basic processes by which caves, from the smallest to the greatest, are formed. Luckily for us, the infinite variety of nature's designs extends underground. In any cave, the exact conditions of development will be quite exclusive to that cave: in some the limestone will be softer, or harder, or thicker, than in others; the water flow will be greater or lesser; the soil cover (and thus the acidity of the water) greater or lesser. All these and other variables result in a great variety of caves, ensuring that the cave visitor will never find repetition.

Cave formations

A vigorous underground stream or deep shaft gives cavers the excitement they seek in their sport. They are also privy to some of nature's most stately and exquisite beauty in the form of cave decorations, usually referred to as formations. It is a beauty hidden far from the light of day, reserved solely for those who would go underground.

Like caves themselves, cave formations are a product of flowing water, but evolve, generally, at a much more sedate pace. As water seeps down through the tiny cracks in the limestone it carries with it the minute quantities of rock it has eaten away. If this water then finds its way into the open air-space of a cave passage it releases some of the carbon dioxide gas dissolved in it — rather like the gassing one sees on opening a bottle of fizzy lemonade. This in turn causes the precipitation of some of the dissolved rock as crystals of calcite.

A drop of this 'charged' water seeping through a cave roof will leave a microscopic ring of calcite around it before it falls. The next drop adds to this, and so on. The result, after many hundreds or thousands of years, is a delicate straw stalactite, which can reach many feet in length. If this hollow straw becomes blocked, the water will flow down the outside, changing its shape into that of the tapering carrot-shaped stalactite.

If the falling drops of water land on rock or clay rather than water, they release yet more calcite, and this deposition builds up the usually much stumpier formation called a stalagmite. Columns are formed when the descending stalactite eventually joins with its ascending companion stalagmite.

Water seeping down a sloping under-cut wall will sometimes produce a hanging curtain formation. Often undulating across their width and translucent to a light shone through them, these curtains can provide some of the most breathtakingly beautiful sights underground.

The most common formation is flowstone – a seepage which can over countless centuries produce whole walls of glistening calcite cataracts, or exquisite frozen rivers winding along the cave floor. Frequently, the water flowing over flowstone will create semi-circular calcite dams, many of them lined with crystals. These dams can vary in size from minute cups to great pools lying right across passages.

Strangest and least common of all are helictites. Springing from the roof or walls, these are the true eccentrics of formations. Their often multiple 'arms' winding and twisting into the passage, they apparently defy gravity — and logic. Cave scientists have proposed a number of possible theories about their weird, contorted growth, but as yet there is no accepted explanation.

All these main formations (there are others) can form either in the pure white of unadulterated calcite or in a whole range of gorgeous colours — creams, browns, reds, greens, blues, yellows, oranges — when the water has been tinted on its way through overlying deposits of minerals.

The combination of exotic shapes and colours, all born of the limestone, creates a true underground wonderland. But remember: these formations are irreplaceable. Never touch them; even if they do not snap (and they do, easily), a muddy fingermark can mar their beauty for all time.

Above Straw stalactites, Cloud Chamber, Dan-yr-Ogof.

Left Stalactites and stalagmites in Swamp Creek, Ogof Ffynnon Ddu.

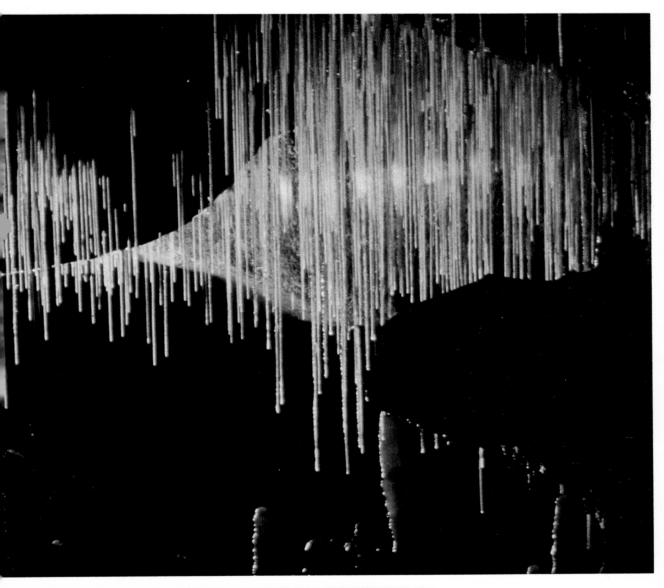

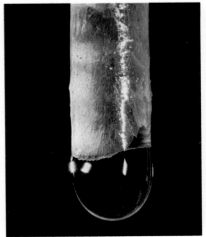

Far Left Tip of a straw stalactite, Ogof Craig-a-Ffynnon.

Left Stalagmite in Ogof Ffynnon Ddu.

Cavers and caving

Underground places in general have always intrigued people – but caves especially, because, as they are natural, we have no way of knowing for certain what lies round the next dark corner. As well as this element of mystery, there is the very real spice of adventure and exploration.

Virtually every square foot of the surface of Britain is known, mapped and photographed. While there is plenty of adventure — on the mountains, moors, rivers and lakes — there can never again be any true geographical exploration. But raise the lid, and underground Britain offers exploration as real and exciting as any unclimbed Himalayan peak.

Physically, caves tax the body to the limit. But it cannot really be pretended that many cavers take up this sport (or activity, hobby, pastime, pursuit – all cavers seem to have their own word) just to get and keep fit.

The real appeal to most cavers is that caves also tax the mind. Caves are graded from 'easy' to 'super-severe', or in rough numerical equivalents, and even during a quick one-hour trip round a low-grade cave the mind attunes itself in a particular way. Coping with the far greater hazards of a long super-severe trip makes far greater demands on the caver's mind: it must attune itself to an environment that is not only alien but also potentially hostile.

This sounds grim, but it simply constitutes the basic elements of challenge in many of man's 'pointless' adventurous pursuits. And, paradoxically, another very appealing point about caving is its apparent pointlessness. It boils down in the end to just you and the cave, and the pleasure derived from extending your own limits of endeavour and endurance, and working right up to those limits – but not beyond!

The number of cavers in Britain can only be estimated, for there is – fortunately – no necessity for registration with, or membership of, any central authority or body. The indications are that between 20,000 and 25,000 people go caving fairly regularly. Although they are spread throughout the whole country, the actual caving regions can be clearly identified.

Right Streamway – Ogof Ffynnon Ddu II.

Cavers and caving

The largest region is in the north of England, chiefly in the Dales of Yorkshire. This is the land of deep shafts and long ultra-demanding caves piercing the very heart of the Pennines. Further down this great backbone of hills and moors is the other major region in the north, the Peak District of Derbyshire, where there is a rough balance between the potholes that predominate in the Dales and the caves of the south. We will look at this difference shortly.

The softer contours of the Mendip Hills in Somerset harbour what is, for its size, one of the busiest caving areas, attracting not only local cavers but many from London too. Mendip cavers have a reputation for sheer determination in their multitude of cave 'digs' — excavations from the surface painstakingly dug, sometimes over many years, in an attempt to find a new cave.

Across the river Severn the great river caves of South Wales lie within the superb scenery of the Brecon Beacons National Park. Cave lengths here are commonly quoted in miles rather than feet, and in many of them long passages echo to the roar of vigorous streams.

Smaller regions are to be found in Devon, Scotland and North Wales, and there is a steady number of new discoveries and extensions each year from the last two in particular. Although outside the direct scope of this book, mention should be made of Ireland's great limestone areas. In regular trips members of English caving clubs have opened up a large number of fine (if flood-prone) caves, and there are now sufficient local cavers to make their own headway with exciting discoveries.

Caves and potholes

It is appropriate at this point to explain the difference between caves and potholes. By and large it is a regional one. In the south of England, in Wales and in Scotland, *caves* predominate — that is, systems that are largely horizontal in development and require no more than standard personal caving equipment to descend. The Yorkshire systems have many more sheer vertical shafts which necessitate special climbing equipment, and take the name pothole or pot. The Peak District falls between the two, both in the type of system and the name chosen.

This is only a rule of thumb, and there are exceptions in every region. As for those who explore underground, while the term 'caver' (and 'cave') is

Above Entrance Chamber to Peak Cavern by S. Prout, published in 1821.

Right Formations in main chamber. Ogof Rhyd Sych.

used as an overall generic name nowadays, cavers in the north of England often prefer to call themselves potholers.

Early cavers

Little is known about the growth of caving in Britain, especially in the early days, when the number of cavers was far fewer than today, and records often scantier. For a start, one has to accept that through the ages a handful of people were always willing to brave the darkness — and no doubt the grim legends — to see what lay beyond the open entrance.

Some of the earliest explorers were lead-miners who, on occasions, broke into natural cavities in pursuit of ore. On Mendip, for instance, miners broke into the 100-foot-high domed chamber in Lamb Leer cave in about 1674; later the system was lost, only to be re-discovered in 1880. And a surprise lay in store for one of Britain's leading cave-divers, Martyn Farr, when after 1430 feet of diving he passed Far Sump in the Derbyshire cave of Peak Cavern in 1981. One can imagine his astonishment when, in a passage he had every reason to suppose would be virgin, he came across signs of previous visits by lead-miners who had made their way in via some now long-lost entrance.

Archaeology, too, led to some early examination of caves, but generally only as far in as remains were to be found; and of course the archaeologist's chief interest lay in what the cave contained rather than in the cave *per se*. Another motivating force over the centuries was commercial rather than scientific.

The first show caves

Unlike today, when almost everyone is a tourist, in days past tourism depended almost solely on the wealthy: only they could afford both the time and the transport. Limestone country offers much in the way of fine vistas, breathtaking gorges and spectacular crags — just the sort of scenery that attracted the tourists of yesteryear as much as today's. Many landowners with caves on their properties saw the potential: after the initial outlay on paths, railings and lamps, there are few 'running costs' in a show cave.

While most modern show caves are run on a very businesslike basis, and some have even attracted investment from large commercial conglomerates, this has by no means always been the case. While at some the visitor would find a fixed fee and a waiting guide, at others he would be charged on a what-they-can-afford basis, given a

Cavers and caving

candle and pointed in the general direction of the entrance!

A number of caves have suffered from half-hearted attempts at commercialization in the past, but now bear witness to these dreams only by the presence of the odd rough-hewn step or two and the occasional rickety railing. Only cavers clatter past these now; no big parties or paying tourists or full car parks. Failure came for a number of reasons: the cave was too remote from other tourist attractions; it was devoid of the formations that other show caves offered; or, most commonly, the money simply ran out.

One of the earliest English accounts of seeing caves as a tourist was given by the Reverend John Hutton (1740–1806) in his book *A Tour to the Caves*, first published in 1780. Towards the end of the 18th century many of the English gentry followed the fashion of visiting the Lake District; some diverted their attention to the caves of the Yorkshire Dales, among them the Reverend Hutton.

Hutton's account is full of reference to the works of Ovid, Virgil, Milton, Dryden and the like. Of being shown through Yordas Cave – his first venture underground – he remarked that:

Several passages out of Ovid's Metamorphosis, Virgil, *and other classics crowded into my mind together. . . . As we advanced within it, and the gloom and horror increased, the den of* Cacus *and the cave of* Poliphemus *came into my mind. . . . The roof was so high, and the bottom and sides so dark, that with all the light we could produce from our candles and torches, we were not able to see the dimensions of this cavern. The light we had seemed only darkness visible, and would serve a timid stranger, alone and ignorant of his situation.*

Caving as a sport

While show caves, archaeology and miner's finds have all served to stimulate interest in caves in the past, they resulted in comparatively little serious cave exploration. The sport of caving – the exploration and study of a cave for its own sake – can be said to have blossomed in Britain from a quite remarkable piece of solo exploration in the closing years of the last century – and that by a Frenchman.

The entrance to the giant Yorkshire pothole of Gaping Ghyll (see pages 32–41) is a huge shaft, swallowing an often vigorous stream. The first attempt to bottom it was made by John Birkbeck of Settle, and it would be satisfying to be able to record his success, but he was only able to reach a ledge at a depth of 190 feet. Most accounts give the date for this as 1872, but the latest research suggests strongly that this was the result of an early misprint, and should probably be 1842. Unquestioned is the date of the 365-foot shaft's eventual bottoming: 1 August 1895, by the visiting French caver Edouard Alfred Martel.

Not unnaturally, this successful descent of such a challenging pothole acted as a strong impetus to the few cavers who, until then, had been pursuing their explorations in an uncoordinated and fragmented manner. It has to be admitted that the fact that it was a foreigner who took the plum undoubtedly worked as a further spur.

Caving clubs

The key to the development of successful cave exploration lay in the birth and burgeoning of the caving clubs. It was the members of northern rambling clubs who did much initial work, such as those of the Yorkshire Ramblers' Club in that county, and the Kyndwr Club in Derbyshire at the turn of the century. On Mendip, in Somerset, it fell to certain members of the Wells Natural History Society to start a more serious exploration of that region's caves. Much impetus was given to caving on Mendip by Dr Ernest Baker, a Somerset-born man who went to live in Derby but often returned to his homeland; this led to a sharing of techniques and experiences. In 1907 he was the principal author of the first book devoted solely to caving, and his daughter was one of the first women cavers in Britain.

Some idea of Baker's prowess is given by another Mendip pioneer, H. E. Balch, in his account of the exploration of Swildon's Hole (pages 60–61), when the now famous Double Pots were first reached.

Far Left Stream Passage in Agen Allwedd just above 3rd Boulder Choke.

Left Organ Chamber, Kents Cavern.

Cavers and caving

We stand looking down into a vertical sided pot-hole, without hand or foothold, with deep water at the bottom, overflowing with a 5ft waterfall into the biggest pot-hole in the cave. . . . The late Dr Baker was the first man to pass these pots, without a rope, putting up his little cairn just beyond. He often climbed where a cat would fall, and we more normal cave-men preferred to take a long rope, and sending one end through the water by an amphibious man, he made it fast to a stalactite mass beyond, so that we above might pull it up tight and secure it to a convenient rock. Thus we could all travel down with our cameras, food-bags, etc., for the deeper cave beyond. The climb back is not so easy.

Dr Baker would, one suspects, be quite at home with the caver of today who climbs the Double Pots with ease and never a rope in sight!

The year 1929 brought the formation of the north's first large club devoted mainly to caving, the Craven Pothole Club. This was followed shortly by the establishment of the Bradford Pothole Club (1933) and, in the south, the Wessex Cave Club (1934) and the Bristol Exploration Club in 1935. The earliest university club – of which there are many nowadays, from Oxbridge to red-brick – was the University of Bristol Speleological Society, dating back to 1919. Until the formation of the South Wales Caving Club in 1946 early exploration in that region was pioneered by members of English caving clubs, and a similar situation prevailed in the great limestone tracts of Ireland until quite recently. Derbyshire has a long caving tradition, but it was only in the 1950s that the first large independent clubs sprang up there.

Both World Wars had a detrimental effect on caving clubs, and in the Second World War at least half, especially in the north, were disbanded. But since then the number of clubs has risen over the decades, the greatest period of growth being in the late 1950s and throughout the 1960s. Today, the number of clubs is probably about 300, ranging from those consisting of just a dozen or so cavers meeting occasionally, to the largest with memberships of 200 or 300 and, usually, permanent club quarters in one of the caving areas.

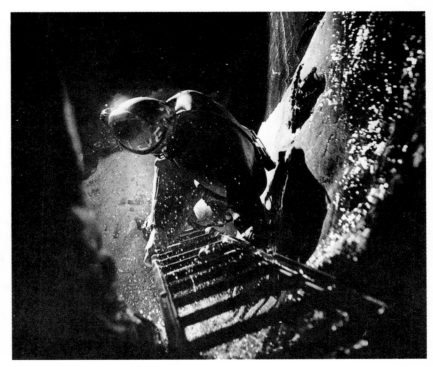

Above Ogof Ffynnon Ddu II – Maypole Inlet.

Left Ogof Ffynnon Ddu – the famous array of the formations known as The Columns.

Cavers and caving

Caving in the 1980s

Apart from the growth of clubs, how else has caving changed over this century? In common with most activities, the hardware of caving has undergone a quite radical transformation. Modern clothing and equipment give a much higher degree of protection than that enjoyed by the pioneers. They dressed in whatever old clothes were to hand, and a wet cave meant a wet and cold trip, with hypothermia a very real threat on the longer forays. Today's electric lamps give good illumination both near and distant; one has to admire the early cavers with their solitary candles stuck into a hat brim, liable to be extinguished − at the most awkward moment, of course − by a single drop of water. Ropes and ladders were a far cry from those used by the modern caver. Without the benefit of nylons and polyesters, ropes were thick and unwieldy constructions of natural fibres, absorbing great quantities of water in wet caves, and liable to rot unless meticulously dried before being stored.

Closely linked with the improvement in equipment, and especially in protective clothing, has come a new breed of caver. Just as in the sport of rock-climbing where there have been incredible advances in the standard of difficulty routinely attempted and reached, so too have standards risen in caving. Many people believe that the 1960s were the watershed years, and it is no coincidence that this is when neoprene wetsuits began to be widely adopted by cavers to combat their greatest problems: the wet and cold of British caves.

Greater protection against the cold probably made less difference to the exploration of the big open caves and pots than to the pushing of the many low, long and constricted crawls that had often previously been left only part-explored. Even with the promise of virgin cave ahead, the hardest caver will find his drive blunted little by little if he has no protection against water. As he shuffles through the streamway the cold is unremitting, for hour after hour, and, along with his body heat, his will is sapped. The limit of previous exploration is reached, as is his own limit. He may push a little further, but he knows that he cannot safely use up as much as half his reserves, leaving half to get out again, for that leaves no margin for the·

Above Ogof Ffynnon Ddu − The Columns.

Right Ogof Craig-a-Ffynnon − pitch below Choke 2.

unexpected. And it is the unexpected which, in this case, can kill.

With better protection and less bulky tackle to carry, the caver was able to push further than ever before. But it would not be the whole truth to ascribe these advances solely to better equipment. The 1960s and 1970s also brought a new breed of caver prepared to push harder than ever before, even taking into account the improved protection against the cold. Climbs were now done free which, before, would have automatically merited ladders, and sumps which once would have been classified as demanding diving equipment, were dived free. There is a clear parallel here with rock-climbing, for over roughly the same period rock-climbers reached higher and higher standards. What were once regarded as artificial-aid climbs only, the climber using metal pegs driven into cracks, are now being climbed quite free. And again, although rock-climbing equipment also improved considerably over this period, this cannot account completely for such a phenomenal rise in standards.

One could look for esoteric psychological or sociological reasons for this increase in the level of risk-taking, and indeed they may be factors, but the author suspects that the reason is much simpler. In both sports − caving and climbing − the most active and experienced participants have found themselves with fewer and fewer *new* delights to entertain them. Gone was the era of roaming the moors and finding open holes yet to be entered, or of walking up to a cliff face and finding many as yet unclimbed routes within one's limits of competence. To experience the pleasure of exploring virgin territory, in or on rock, both caver and climber had to work to ever higher, harder, standards.

In the north, as elsewhere, caves and pots were driven even deeper where humanly possible. Quaking Pot provides just one such example. First explored in 1942 by the British Speleological Association, it was listed in the old (1959) guidebook as 321 feet long and 180 feet deep − and even then had the ultimate grading of Super Severe Pot. Today's guidebook reflects the greater determination of members of the prestigious University of Leeds Speleological Association in 1964, and, nine years later, that of the Preston

Cavers and caving

Caving Club. Now the pot stands at 1870 feet long and is 467 feet deep. A clear indication of how hard it is is given by the guidebook's introduction to the route description:

WARNING – A serious undertaking. Strenuous moving tackle from Crux to sixth pitch. Crux is more awkward for tired people on the return and rescue from beyond it would be virtually impossible.

In Wales, a good example of this new, more aggressive, approach to pushing was provided by the extension of a certain passage in the large system of Agen Allwedd (pages 52–53). Southern Stream Passage was initially followed for only part of its full length, then discounted, in cavers' parlance, as 'not going'. For years its reputation kept cavers at bay: it was very low, very uncomfortable, and wet. It was a small caving party of the new breed who refused to accept that it was choked irrevocably, crawled along its twisting gut, and, by merely moving a few rocks, opened up the full 6000 feet of passage to the cave's fine Lower Main Stream Passage and bottom sumps.

As well as under their home ground, British cavers have also turned their exploratory drive to good effect in cave systems in ever more remote corners of the world. New standards have been set by highly experienced British teams mounting large expeditions to limestone areas as far away as Sarawak and Papua New Guinea.

The sport in practice

Today, as through the years, there are a number of cavers who prefer to follow their sport in the most informal way possible. When they find a free weekend to hand, a few telephone calls will suffice to locate three or four likeminded friends. A rendezvous is fixed, each caver turns up with his share of the tackle, if any is needed, and a caving trip is under way. Most cavers, though, much prefer to join an active club which meets regularly and within reasonable travelling distance of their home.

Membership of a club has several advantages for a caver. There is the purely practical one of having access to shared facilities beyond the scope of the individual – large libraries, accommodation in a caving region (ranging from stark barns to purpose-built headquarters with dormitories, showers, kitchens, lamp-charging units, etc.), and, in these days of expensive petrol, shared transport. Modern caving ladders and ropes are not cheap, yet when their cost is spread across a club's membership the caver has access to all the tackle he may require without any great expenditure.

For the beginner, joining a good club with a full meets list is an excellent way of learning the game. He or she will be steered on to the meets that match their experience, and shown informally the best and safest ways of surmounting each cave's difficulties. Such skills can be imparted at the various training courses that are available (such as those at Whernside Manor – see page 29), of course. But the particular bonus that comes from learning as a member of a club is the friendships that are struck up as a consequence.

Left The Canal Bypass, Little Neath River Cave; the route to be taken when the water is high.

Below Climbing a cascade in the stream passages in Ogof Ffynnon Ddu.

Cavers and caving

For as well as all these practical advantages, the successful club caters for the close companionship of cavers too. While part of caving's pleasure lies in being able to cope as an individual, there is the shared pleasure of tackling difficulties and dangers as one of a team. This is reflected on the surface in the social life of cavers with the evening's session in the pub, sharing tales of the day's trip. In such an intimate atmosphere, clubs flourish.

While membership of a club takes care of the individual caver's immediate sporting and social requirements, there are larger issues to be considered. Chief among these, as with any similar adventure pursuit, is rescue.

Rescue and access

The firm establishment of the club structure is reflected in the organization of cave rescue. There is no national body to which rescuers must belong, simply regional organizations (e.g. the Cave Rescue Organization, the Mendip Rescue Organization) which initiate and co-ordinate rescues. All those involved in rescues are volunteers, and there is no outside official funding – or interference. A parallel could usefully be drawn with the lifeboat service. The over-riding theme is one of self-help, and cavers are happy to keep it that way.

Not unexpectedly, in view of the individualistic nature of their sport, cavers on the whole dislike any sort of outside or 'official' interference. Unfortunately, sport, in common with other aspects of society, has had to accept a growing amount of bureaucracy, and caving has been no exception. In the 1960s, northern cavers found it necessary to set up the Council of Northern Caving Clubs, chiefly to deal with questions of access to certain caving areas, and their counterparts in the south followed quickly with the Council of Southern Caving Clubs. Other regions followed suit. The next step was the establishment of the National Caving Association, and this continues to operate on the basis of regional representation by the various councils together with certain specialist bodies. The significant point is that the NCA was formed as a representative body and not (as in some other countries) to control or regulate the activities of cavers in any way.

Access to caves can raise a number of problems, and in many cases, in a typically British way, is subject to compromises. Fortunately, there is still a reasonable number of caves to which the landowner will allow free access, often for a small goodwill fee. In other cases (perhaps where there have been problems in the past), clubs or regional councils have taken over control of access from the landowner. With some caves – especially in the Dales – limits are placed on the number of parties or cavers allowed down on any one day, and very popular systems can be booked up for many months ahead.

Conservation

Keeping access to caves open is one problem. Cave conservation is another major one. Sadly, each year irreplaceable formations are damaged or even removed from caves. Sometimes, of course, the damage is completely accidental, but a very small but persistent number of people who go underground seems to be blind to the fragile nature of the cave environment. Short of sealing all entrances for good, the only potentially effective solution would appear to be constant education about the need for conscious conservation. Again, the club structure is best placed to handle this at the grass roots level.

Caving equipment

Deep inside a cave, man is in a distinctly alien environment. Often he is crawling or even swimming through bitterly cold water whose sole purpose seems to be to suck every last iota of warmth from his body. The rock around him is hard and unyielding, and has a habit of suddenly disappearing from underfoot at some sheer plummet. Living off the land is out of the question – and someone has turned all the lights out.

The caver, then, needs all the extra help he can get, and today he can call on much more than his predecessors could in the way of clothing and equipment. Until well into the late 1960s, the British caver relied largely on the cast-offs of the armed forces and coal miners, with a little help from the rock-climbing fraternity. Then certain manufacturers recognized the existence of a market crying out to be sold to, and specialized equipment started to be made available. The result? The caver of today has at his disposal a range of goods which the pioneers would have gasped – and no doubt grasped – at.

As this book is in no sense a 'how to do it' guide, a very detailed description of modern equipment would be inappropriate. But for the reader inquisitive as to how the caver *does* cope with that alien environment – what equipment he has at his disposal; what special techniques he employs to surmount the many different obstacles and hazards – the following will provide a good outline.

Head protection has improved progressively over the years. Early cavers had little chance of coming off best in a collision with a cave roof when only a flat cap or soft hat lay between the two. Miners' fibre helmets have brought two advantages: protection against knocks, and somewhere to hang your light. The latest glass-fibre helmets also give protection if the caver falls.

Candles have been usurped by acetylene lamps, and the simple helmet-mounted type has remained basically unchanged because of its inherent simplicity and (when its individual quirks are appreciated) reliability. The top half is a small water reservoir which drips, at a controlled rate, on to lumps of calcium carbide held in the bottom container. The interaction produces acetylene gas which burns at a small jet in front of a polished reflector. However, except on long expeditions where mains electricity is rather remote, rechargeable electric miners' lamps are increasingly favoured. These give a powerful beam of light for up to twenty hours, and work in waterfalls or even underwater.

Perhaps the most significant development has been the use of wetsuits, like those worn by divers. Trapping a layer of water next to the skin, and holding it at near body heat, the foam-rubber wetsuit has greatly extended the cavers' ability to withstand long periods of immersion in cold water. This has contributed significantly to the extension of extremely hard, long and wet systems.

Water, and the attendant threat of exposure (hypothermia), is perhaps the greatest and most frequent risk. Cavers pushing through a precarious boulder choke – where fallen boulders almost fill the passage – or climbing up, over or past some obstacle take a calculated risk. Yet, let one merely twist an ankle

Right The breathtaking majesty of the 175ft final shaft, Juniper Gulf.

Cavers and caving

deep in a hard, wet cave, and the greatest risk is that exposure will lay claim before one can be got out to safety.

With helmet, light, boots and – in a wet cave – wetsuit, what else does the caver need? Theoretically, nothing at all, but the wise caver will guard against mishaps by taking along a spare source of light; basic first-aid kit; pencil and paper; spare carbide and water or electric bulb; and emergency food. On a long trip, suitable high-energy food is also essential.

Caving techniques

Horizontal caving

Most cavers spend most of their time involved in horizontal caving: working their way down, or out of, a cave through more or less level passages. While show caves give a good idea of the variety of larger passage profiles, by their very nature they cannot show the common types of caving obstacle.

The height of the passage dictates how the caver moves. Given a flat floor, free of boulders, and a roof from 7 to 100 feet or more above, he can enjoy the luxury of walking, with no more difficulty than along a pavement. But this is usually shortlived, alas. Down comes the roof, and the caver moves forward in a stoop. Down a little more, then even more, and progress turns first to a hands-and-knees crawl, then a flat-out crawl. Nature usually contrives to have a pool or stream in these lowest sections, and, while 6 inches of water would barely be noticed elsewhere, in a 1-foot-high crawl it does tend to intrude.

The fear that springs to the minds of most non-cavers when they comment on the sport is that of getting stuck. In truth, very few people get seriously stuck for any length of time in caves. It is surprising just how small a gap an experienced caver can squeeze through, given the right technique – which is vital – and coolness of mind. Many apertures which, reconstructed on the surface, the onlooker would class as impassable, are run-of-the-mill constrictions for the caver.

The greatest risk of injury is probably during free-climbing, or clambering. Even though technically horizontal, many passages force the caver to scramble at a high level, or clamber up and over obstacles perhaps 10 to 15 feet high. Such obstacles rarely require a lifeline and handline, yet a minor lapse

of concentration can lead to a fall great enough to cause a bad sprain or even a breakage.

As well as length, every cave has depth. In some caves depth is achieved by passages gently sloping over a considerable distance. In others, it is achieved dramatically.

Vertical caving

Occasionally, vertical shafts (or pitches) have sufficient holds to be climbed free. Usually they require rigging for the descent and ascent.

The traditional method of rigging, still widely followed, uses electron ladders. These were developed from rope-ladders, and have steel-wire sides (as opposed to the old hemp rope) and aluminium-alloy rungs just wide enough to accommodate one boot. They are both lightweight and strong, but do demand a particular climbing technique: the climber's bodyweight must not be placed unduly on the arms as the ladder will swing away.

On all but the shortest ladder pitches, the climber is secured on a lifeline – of nylon or other artificial fibre – paid out by a lifeliner who is belayed (tied) at the top of the pitch. When it is the lifeliner's turn to climb down, the doubled rope is clipped into a pulley and is secured by one of the cavers at the pitch bottom.

Since the early 1970s, an alternative method of climbing pitches has come into favour: single rope techniques – SRT. As the name might suggest, this involves rigging a pitch with just a single rope. Not only is the length of rope needed halved, but the necessity for bulky and heavy ladders is entirely eliminated – a vital factor in a speedy trip down a deep cave with a small team.

For the descent, cavers use their own preferred make of descender: a metal device that controls the rate of descent down the rope, from which the caver hangs in a tape harness.

When the journey back to the surface has to be made, cavers use several devices known as ascenders or prusikers. The caver stands in tape loops attached to the ascenders and slides them, one at a time, up the main rope. Although they slide upwards easily, when the tape loops apply downwards pressure they lock on to the rope. By sliding up one at a time, cavers create their own moving ladder, and can 'walk' up the rope.

Because of the risk of entanglement with the main rope, no lifeline is used

with SRT. This entails absolute reliance on the main rope, and to prevent damage by abrasion protective pads are used where the rope passes over rock edges.

Particularly with a large team, it is usually quicker to rig a small pitch – say up to 50 feet – with ladders, saving the time it takes each caver to assemble SRT equipment on the main rope. On longer pitches (up to 400 feet), SRT comes into its own as the more efficient method.

Water obstacles

Water is a constant problem in many caves. Apart from the hazard of exposure, it presents cavers with two major obstacles – and only inches separate them!

Where the cave roof comes to within a few inches of a stream, a duck confronts the caver. Depending on exactly how few inches of airspace are left, the helmet may even have to be removed before the caver slides into the duck, which stretches ahead for some yards. The head is tilted to one side so that the nose and mouth can just reach the tiny amount of air. At times like these, one doesn't make waves.

Of course, if the roof is a few inches lower, it will actually dip below the water surface. This daunting water-trap goes by the name of a sump, and whether it can be passed by free-diving (i.e. without diving equipment) or not depends on its length and the size of the submerged passage. Hand-lines are laid through such places for the caver to follow, but this still does little to remove the stark challenge of holding one's breath and diving into a black submerged tube, knowing that the only way up to air is at the far end.

The specialists

After some years of straightforward caving, many cavers turn to a specialized branch of their sport. One of the most popular, although by no means the simplest if first-class results are desired, is cave photography. Cave photographers have the great advantage of being able to control *all* the lighting at their disposal, but have to deal with the enormous problems of

Right The 1st pitch in P8 (Jackpot), Derbyshire.

keeping their equipment dry and clean and ensuring that the flashes go off when they are supposed to.

Exploring caves can involve specialists such as rock climbers and, more and more often these days, cave-divers. The latter is a super-specialist, pushing risk to the very limit in the effort to reach virgin cave by passing sumps many hundreds of feet long.

There is also a welter of purely scientific disciplines, ranging from the study of how and when caves were formed to the observation of the life forms that eke out a precisely balanced life style in the environs of caves.

Starting caving

Anyone interested in going caving is strongly advised to do so in the company of an experienced caver. One course is to make contact with a local caving club, of which there are many spread throughout Britain. These are listed in the *Descent Handbook*; details from Mendip Publishing, Cleeve House, Theale, near Wedmore, Somerset, tel: 0934–712995. Advice can also be obtained from the National Caving Association, c/o Whernside Cave & Fell Centre, Dent, Sedbergh, Cumbria. Alternatively, you could attend one of the beginner's courses at Whernside; full details obtainable from the address above.

WILD CAVES

1. Gaping Ghyll
2. Juniper Gulf
3. Lancaster-Easegill System
4. Meregill Hole
5. Penyghent Pot
6. Providence Pot to Dow Cave
7. Giants Hole
8. Jackpot (P8)
9. Nettle Pot
10. Agen Allwedd
11. Ogof Craig-a-Ffynnon
12. Ogof Ffynnon Ddu
13. Little Neath River Cave
14. Swildon's Hole
15. St Cuthbert's Swallet
16. Eastwater Cavern
17. Uamh an Claonaite

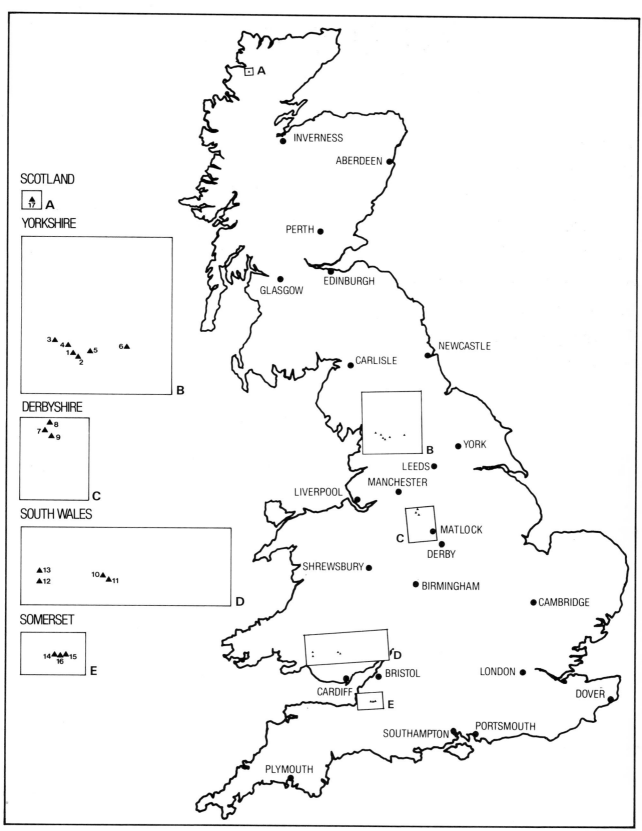

SCOTLAND

YORKSHIRE

DERBYSHIRE

SOUTH WALES

SOMERSET

INVERNESS

ABERDEEN

PERTH

GLASGOW

EDINBURGH

NEWCASTLE

CARLISLE

YORK

LEEDS

MANCHESTER

LIVERPOOL

MATLOCK

DERBY

SHREWSBURY

BIRMINGHAM

CAMBRIDGE

BRISTOL

CARDIFF

LONDON

DOVER

SOUTHAMPTON

PORTSMOUTH

PLYMOUTH

Gaping Ghyll

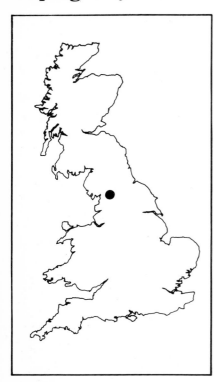

A cave is a book telling two stories. The pages are composed of countless layers of limestone, themselves faithfully recording the deposition, millions of years ago, of sand, silt and a myriad of minute skeletons of tiny sea creatures drifting down to the cold sea bottom from the warm zones in which they once lived. Through these pages runs the second story, that of the slow growth of the cave itself, with the whole history imprinted on the virgin pages not by ink but by the passage of water.

Correspondingly, to explore a cave is to read the book, page by page, chapter by chapter. The story of Gaping Ghyll is one which cavers have been reading avidly for the whole of this century, and, in the best tradition, it has teased out its surprises and revelations all the way through.

Even cavers who have yet to visit Gaping Ghyll (or Gill) will refer to it in near-reverential terms, for they know that this — the grandfather of British caves — is a magnificent and complex system of great variety. Even if it had been discovered only recently it would still stand as one of our greatest caves. But an added degree of importance is imparted by the fact that its first descent marked the birth of serious organized caving in Britain. It seems fitting that its own history of exploration has been one

of steady progress through the century, for so many of our great Yorkshire caves and potholes revealed virtually all their secrets in the earliest days. Since cavers were first privileged to witness its great halls and byways, Gaping Ghyll has presented as many fresh mysteries as have been solved, step by step.

Above the small pretty village of Clapham, one of the popular 'gateways' into the 680 square miles of the Yorkshire Dales National Park, rises the great crouching bulk of Ingleborough mountain. From its blunt summit at 2372 feet the sides of the main bastion on this approach fall steeply enough, but then relent a little before a final steepening again for the drop to the village. The slopes of this final flank are cut through by the pleasantly wooded valley of Clapdale (now a nature trail) with Clapham Beck gurgling its way down to the cottages below. A mile up the dale from the village, at the bottom of a 60-foot high cliff on one bank, is the entrance to Ingleborough Cave, a show cave which has also answered at various times to Clapdale Great Cave and, more recently, Clapham Cave. Close by this entrance, at Clapham Beck Head, the waters of the beck flow into the daylight, having taken a separate ¼-mile route from the innermost reaches of Ingleborough Cave.

The resurgence of the stream at Ingleborough Cave is not merely a place of passing interest for the caver *en route* to greater things. The full story of a cave does not end until the final re-emergence of its stream or river at the surface. Sometimes, for the caver, the pages of this final chapter remain uncut, utterly inaccessible for even the cave-diver. The caver must be content with watching the cold waters spring to the surface from fissures too small to enter. But throughout this century the thousands upon thousands of cavers who have tramped past Ingleborough Cave have felt within them a bright dream rekindled, for they have known that these waters come from no less a source than Gaping Ghyll itself. Knowing that the resurgence is part of an

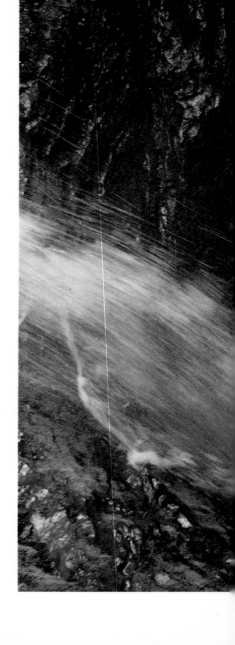

Right Geoff Crossley abseils into the Main Shaft of Gaping Ghyll.

Gaping Ghyll

open cave, the caver's great dream has been to link the two caves, one day to enter the great cave in the mountain above and re-emerge at the cave mouth in this cool dale. We shall return to this dream later.

With the final short dog-leg through Trow Gill, the sides of the dale, which now points to the summit of the mountain, diminish, vanish. Ahead, on the steep flanks of Ingleborough and its lower outrider, Simon Fell, scattered springs add their quotas to the water of Fell Beck. But the gradual descent of this beck lasts for only a short distance, for, without warning, the water finds itself trapped in the steep sides of a long shakehole. There is only one way for it to go: the sudden and dramatic plunge into the maw of the Main Shaft of Gaping Ghyll.

This great gash, piercing the limestone to a depth of 360 feet from the level of the surrounding moor, has been a source of mystification and wonder for centuries. On the hot days of summer the cool air of the shaft mingles with the warmer atmosphere of the surface to produce a mist, and when Fell Beck roars in full flood the thundering water, crashing to the darkness, causes wraiths of fine spray to eddy round the slopes of the shakehole.

The story of how the challenge of this huge shaft was met is a fascinating one. It was first taken up in the last century by John Birkbeck of Settle, 7 miles down the road from Clapham. His plans were very thorough. In what must have been an expensive operation, he had a trench dug to divert the waters of Fell Beck to another sink, giving him a fighting chance in the absence of the great waterfall – the highest in Britain. It appears that Birkbeck made two descents, assistants lowering him on the end of a rope. The first very nearly proved fatal when strands of his rope were severed on a rock edge, but on the second he reached the ledge which now bears his name, 190 feet down the shaft. This ledge also marked the limit of Alfred Clibborn's descent in 1882, again on the end of a rope.

No doubt drawing lessons from the experiences of his predecessors, Edward Calvert – one of the founder members of the famous Yorkshire Ramblers Club – was nearing the end of his own preparations by 1895. Realizing the drawbacks of being lowered on a rope, he planned to use rope ladders in his attempt, which he was busy manufacturing with the help of friends. Other moves were afoot, though, which were to destroy Calvert's aspirations.

In the summer of that same year, a great pile of rope ladders and hundreds of feet of rope arrived in Clapham, the travelling luggage of the renowned French caver Edouard Alfred Martel, in England to address a geographical congress – and have a crack at the great pothole. He had the full co-operation of the owner of the Ingleborough Estate, and thanks to this Birkbeck's old trench was refurbished and extended. On 1 August, before a crowd eager to witness the Frenchman's attempt on the shaft, Martel knotted together his lengths of ladder and lowered them into the darkness.

Considering that the telephone had been patented by Bell barely 20 years earlier, it says much for Martel's innovative approach that he relied on this instrument to communicate with the surface, where his wife would translate his instructions to the eager gang of volunteer lifeliners.

Despite all the efforts put into diverting Fell Beck, enough water still plunged down the shaft to soak Martel to the skin. At 130 feet down, he was forced to hang in the full freezing flow until a knot in his lifeline could be released from a crack. The rope freed, he reached Birkbeck's ledge to find the ladders piled in a heap. These he threw clear, the crashing noise causing a few missed heartbeats on the surface. The descent continued, Martel repeatedly swinging into the waterfall only to be buffeted out again. Less than half an hour after leaving the surface, the Frenchman had reached the floor of Gaping Ghyll's Main Chamber, the first person ever to stand in awe before the full majesty of this, the most impressive chasm in Britain.

Almost 500 feet long and 90 feet across its width, this chamber soars craggily to a roof up to 130 feet above the boulder-peppered floor. With only their thin lamp beams, cavers seeing it for the first time at night can hardly grasp at the great dimensions. Only by day does one experience, like Martel, the full impact of the sight. For then a huge column of light, varying from opalescent grey to scintillating silver according to the degree of sunlight playing on the surface, rises from the floor in the middle of the chamber to be swallowed by the shaft piercing the giant arching span of the ceiling. Down the middle of this column of light rides the waterfall, not a solid bore but a con- stantly undulating flow fractured by the great fall into a million separate spears of water. When Fell Beck growls in flood, more than half a dozen individ- ual columns of water crash down the shaft into the flooding chamber, whip- ping up a mass of spray swirled away in countless separate gusts of air. At these times the observer appreciates the appropriateness of the chamber's other name: the Hall of Winds.

Martel spent over an hour exploring the Main Chamber, sketching it to a surprising degree of accuracy consider- ing the extreme pressure under which he was working. Had he or his ladder suf- fered an accident, rescue would have been most unlikely. As it was, despite the rather over-eager pulling of his life- liners, and the severing of his telephone wire, he reached the surface safely. Quickly, the news of his discovery spread. The seed sown by Birkbeck had taken firm root, 360 feet deep in the Pennines, and from Martel's triumph the growth of organized caving in Britain can be traced.

When Edward Calvert became the first Englishman to touch the floor of the Main Chamber, ten months after Martel, he also advanced the sport in an important way. Realizing that the descent and ascent of 360 feet of rope ladder would leave little spare energy for further serious exploration, he made his descent on a seat lowered by a windlassed rope. This technique, even- tually refined over the years to use a motorized winch with steel cable instead of rope, made it possible for cavers to drop into the heart of GG in relative comfort, leaving them fresh for the job in hand, the unravelling of the system's secrets.

Although the use of a winch makes for an easy entrance into the cave, the elaborate setting-up of such a system rules out its use on a weekend by week- end basis. What was really needed was another way in, one involving less dramatic ladder descents than in the great Main Shaft. In the event, not one but four additional entrances were dis- covered: Flood Entrance Pot, Dis- appointment Pot, Stream Passage Pot and Bar Pot.

Right The cathedral-like grandeur of the Main Chamber.

Gaping Ghyll

Flood Entrance Pot was the first to be found, in 1908, and provided a sporting alternative way in. There is relatively little horizontal passage to be negotiated, and a series of short ladder climbs leads to the excellent last pitch. This is 140 feet deep, and at its foot continues the gaping hole of South-East Pot, another 130 feet deep. The descending caver must pendulum on the end of the ladder to gain a footing on his objective, South-East Passage, and the last caver down has to remember not to let the ladder swing back out of reach of the party.

Stream Passage Pot was a much later find, by the Northern Pennine Club in 1949, and it is a fine example of a sporting, and still forming, classic Yorkshire pot. An entertaining squeeze down through boulders packed at the entrance leads to the first pitch of 25 feet, through hard black glistening limestone, then to 600 feet of walking-size streamway. Three further superb pitches follow (of 90, 110 and 75 feet), the last two following one end of an enormous rift, and all are savoured by those who appreciate wet pitches – especially when rain clouds have recently swept over the Pennines. This entrance lands the caver in the far western limb of Gaping Ghyll, in Stream Passage, within easy striking distance of the Main Chamber and the host of other passages that trend in their development towards the south-west.

A detailed description of all 7 miles of Gaping Ghyll would be both pointlessly confusing and beyond the scope of this chapter. Instead, let us follow the example of many cavers visiting GG and trace a typical cross-section through the labyrinth by undertaking a through-trip. Many caves and potholes are straight in and out, down and up affairs, but those such as Gaping Ghyll, with at least two quite separate entrances, give large groups of cavers the opportunity of a long trip with no retracing of footsteps. The group splits into two teams, each tackling its chosen entrance and exiting via the other, detackling as it does so.

In the shorthand of cavers' pub talk, our particular through-trip would be described as Bar-Hensler's-Dis – hardly informative to the outsider, but to those who know Gaping Ghyll the formula for many hours of hard, varied and satisfying caving.

Four hundred yards to the south of the shakehole harbouring the main shaft is a very large depression. The entrance to Bar Pot is an obvious hole at the base of the highest of the rock faces which line the shakehole. Short free-climbs lead down to the head of the first pitch, where a 55-foot ladder is belayed to a fixed iron bar. At first the climb is within a narrow rift, but this blossoms open into a chamber with the ladder landing on a bed of boulders. At the bottom of the chamber a hole leads through to a descent down a smooth and steep slab requiring some cautious climbing.

At the foot of the steep boulder slope one can continue on either side, but we divert to the left, along a muddy passage, then under a large rock bridge to the head of the big pitch. This drop of over 100 feet – dry, and close to the wall – has probably initiated more cavers into the 'Hundred Foot Club' than any other in the country, and, being devoid of nasty surprises, it is ideal for this purpose. The landing is on the level and dry floor of Far South-East Passage, bringing one painlessly to the ground floor of this fascinating system, relatively fresh for the challenges ahead.

To the right (as one faces the ladder), Far South-East Passage has a limited career, coming to an abrupt choked end after only 600 feet. It should be mentioned, for completeness, that this passage is not utterly devoid of continuation. In fact, the small hole at floor level not far along gives relatively easy and quick access to a part of the cave we shall be reaching later, but via a much harder route. The kindest thing is to forget this short cut. Our way is to the left.

Very soon, at the point where Flood Entrance Pot pierces its own independent way into the system, there is a cautious traverse past the waiting mouth of South-East Pot, dropping

Left Abseiling down the magnificent Main Shaft.

Right Northgate, in the Far Country Section of the cave, deep under Ingleborough Mountain.

Gaping Ghyll

sheer and hungrily 130 feet to the level of the water table, a region where cavers hope for big discoveries in years to come. Much of the next 400 feet, along South-East Passage, is at that most awkward and annoying height: too low to stoop, but high enough to make crawling *hardly* necessary. But, as always, the ceiling dictates your stance: you crawl, impatient for the next junction. At this, our chosen route diverts sharply right into South Passage.

Straight ahead, though, and worth the short detour, lies Sand Cavern, an impressive place some 300 feet long and 50 feet wide, with its roof – up to 30 feet high – pierced by solution tubes. At the far end is a great wall of stratified sandy clay. When the original explorers passed it in 1905 by climbing to a sandy crawl almost at roof level, they found exquisite stalactite straws.

From the junction mentioned earlier, the arched South Passage gives easy progress now, past the small Pool Chamber and the constricted Portcullis. Branching right is the low sandy-floored Booth-Parsons Crawl, of which more presently. The last few feet of South Passage are negotiated in silence; all prepare themselves for the spectacle to come; even newcomers sense the air of expectancy.

Suddenly, there it is, the breathtaking chasm which is the Main Chamber of Gaping Ghyll. You have entered at a point on the south wall, and the Hall of Winds spreads to the left, to the right, ahead – and ever upwards to that great ceiling. It is a profound moment, and the statement that the chamber could virtually swallow York Minster seems utterly apt. The furthest reaches lie away in the gloom of distance. The irresistible focal point is that column of light with its foaming, fuming core of falling water, Fell Beck in 360-foot flight. Quickly the water is absorbed between the countless boulders of the level floor, taking passages as yet unseen by man, until, that is, it emerges in the innermost reaches of Ingleborough Cave.

Someone is falling! A tiny figure, headlamp a mere glowing pin-prick, is falling down the Main Shaft surely – but the descent is unaccountably at a slight angle from the perpendicular. The figure, only yards from the floor, slows rapidly, like some aerial flyer in an exotic pantomime, descends the last few feet at a safe speed. Another visitor has just experienced the stomach-lurching winch descent into GG, much of it

virtually free-fall, the boson's chair being pulled clear of the worst of the waterfall by a fixed guidewire. On two weeks of the year, two Yorkshire clubs take it in turns to organize a winch meet, and the electric thrill of this easiest of all ways into the Main Chamber can be experienced by cavers and other visitors alike. Taking advantage of this infrequent chance, many cavers camp on the surface close to the entrance; 40 or 50 tents can spring up, creating a small remote community with one aim in mind, to probe ever further into the complex beneath, more than 7 miles long.

As Martel suspected after his descent, there are indeed ways through the great boulder piles at each end of the Main Chamber. That to the west merely loops back to South Passage, but the Old East Passage takes us through a once beautifully decorated cave (now much less so – another sad example of the need for careful conservation) to the huge Mud Hall, the second largest chamber in the cave. Because of its liberal coating of mud (which streams are now steadily removing), it is difficult to appreciate the true size of Mud Hall without the benefit of daylight. Its scale is realized most readily when a caver has climbed the enormous final boulder choke to the passage's continuation, and his bobbing light gives some idea of the distance.

Old East Passage continues straight from Mud Hall to a constriction so tight that, although a light in the passages of Carr Pot could be seen beyond, not even the slimmest caver can pass through. And cavers have decided to leave it thus, for blasting open the constriction would provide an all-too-easy route into the North Craven Passage and its magnificent but ever vulnerable formations, at present only reachable by a complete descent of the difficult Carr Pot. Shortly before this tantalizing barrier, though, parties can turn right into Far East Passage and continue on past delectable straws many feet long to Whitsun Series, discovered after much work in the late 1960s.

Such are the delights waiting beyond Mud Hall. Something else is in store for our party, however. Just before South Passage burst into the Main Chamber, a low passage branched off: the Booth-Parsons Crawl. This is followed for just over 100 feet to where a low uninviting slot leads off into the blackness. Those who have gone through this slot before wonder what on earth has brought them

here again; those who have not wonder if all those horror stories are true after all. They are.

This is the start of the notorious Old Hensler's Crawl, a test not of speed or polished climbing technique, but of sheer slogging endurance. The crawl is over a ¼ mile long – a fact most try to forget once they are grunting their painful way along it – and along its entirety very rarely does the ceiling exceed a height of 18 inches. At its lowest, the floor and ceiling are close enough to jam one's boots if one tries to roll over to a less painful crawling position. Were this all, it would be quite hard enough. This is not all. Frequent pools stretch across the entire width, and although only a matter of inches deep they soak the crawling caver completely, and are bitterly cold. The finishing touch is provided by the nature of the floor across which the caver drags himself: the hard unyielding stone has been raised into a million ripple ridges each an added point of pain for a too-hastily placed elbow, hip or knee. Flood debris caught on parts of the ceiling adds to the general atmosphere of gloom.

With each foot covered, the caver's admiration grows for Eric Hensler, the Londoner who, quite alone, discovered and explored this long hard gut in 1937. Not for him the protection of a wetsuit, nor the encouragement of other cavers. Hensler was eventually rewarded for his remarkable solo effort by the discovery of Hensler's Stream Passage. Here the water has broken free of the constraints of the crawl to carve out a large canyon which swings south-east to head promisingly in the rough direction of Ingleborough Cave's inner reaches. Soon, however, it forks into two parallel continuations, both ending in dark and ominous sumps. These marked the end of progress this way for 30 years. Then, in the same productive year that the Whitsun Series was discovered, an iron ladder was pushed up into a muddy tube above the sumps and a 60-foot shaft descended to complete the sump bypass. Beyond, jealously guarded by the flood-sensitive Southgate duck, with only inches of air-space in normal water conditions, lie some of the most remote and demanding passages in Gaping

Right A busy weekend with three separate ladders down Bar Pot, the most popular entrance to Gaping Ghyll.

Gaping Ghyll

Ghyll, the 1¾ miles of Far Country and Far Waters. Both rewarded their discoverers with fine formations, but, even more important, the extremities of Far Country took them very much closer to the dreamed-of link with Ingleborough Cave.

However, the demands of a trip into such territory are not best met by a full and tiring traverse of Old Hensler's Crawl beforehand. Cavers heading there will take the easier approach route via New Hensler's Crawl (that short-cut mentioned earlier, leading off from near the bottom of Bar Pot's main pitch) or the glutinous Mud Hensler's Crawl from Mud Hall. We, enjoying the luxury of standing upright again at the exit from the long crawl, have a different objective – Disappointment Pot, and the start of this alternative entrance into (or exit from) Gaping Ghyll lies only a few minutes' easy going away.

Disappointment stands in sharp contrast with Bar Pot, the route by which we entered. While the latter is dry and rather dead in character, Disappointment is a fine, clean and active pothole, as wet and sporting as any caver could require in its pitches and 600 yards of streamway.

A clamber up a very large boulder slope, and there is the foot of the ladder (left, as all the others will have been, by the other half of our exchange trip party, or perhaps by the club responsible for running the week's winch meet) beckoning us to leave the Hensler's series for higher and more exhilarating things.

We start the climb which, though only 45 feet high, is the longest in Disappointment – another bonus. The ascent is against one wall of a very large chamber which soars giddily, and the next ladder climb, of 25 feet, takes one even higher up this wall, to where the water crashes in from a passage, our exit.

The passage closes down, the trench cut into the floor diminishes, and it is back to the business of crawling once more through a bedding plane. Not for long, though, for soon the foot of another pitch is met, the stream taking the 30-foot shaft in its noisy stride. Now the party works its way along the close confines of a very long high-walled canal, sometimes walking, sometimes shuffling sideways in crab-manner, occasionally crawling where the sides pinch in awkwardly.

Past twin avens piercing sharply into

the stone above and sudden zig-zag bends in the canal, we face the last two ladder climbs – 25 and 35 feet – each requiring at the top a careful traverse exit along a ledge. Again the passage roof lowers, this time with a steady earnestness which betokens some final sting in the tail. And sting there is, in the shape of a low aquatic crawl with, as the crux, a duck offering only the barest breathing space. With helmet off and pushed ahead in one cold numbed hand, you twist your head so that at least one nostril is clear of the stream's surface. The duck is not very long, but no one makes waves!

Beyond the duck, as though with a final flourish, the streamway twists and winds for its last 100 feet. A hand-over-hand clamber up a rope fixed at a steep slope, a last exertion up a 6-foot climb, and you flop thankfully on to the grass of the small shakehole. The others, talking excitedly with that release of tension that comes at the end of any big caving trip, join you. Someone remarks wryly that the hole you all went down those long hard hours ago lies a mere 300 yards away over the rough stubbly Pennine grass. But though these two points mark the beginning and the end, the journey over those hours has been a unique and unforgettable one through the ancient hills and waterways of one of Britain's finest cave systems. Its magic has been described by a keen caver:

Night-old stone, neatly incised,
Kindles dreams,
And deep nature allows some –
Neophyte interlopers, laughing, lithe –
Old caver's knowledge;
Cold, under, forever.

What of that dream – of a caver descending Gaping Ghyll and surfacing through the lower sister system of Ingleborough Cave? One of the greatest delights of British cavers of this era is that this is a dream no more, but exuberantly celebrated reality.

From Martel's descent of the Main Shaft, and the first explorations of Ingleborough Cave in the last century, cavers have been piecing the giant system together, foot by foot, overcoming enormous obstacles and dangers. Ultimately it fell to the cave-divers to make the final connection through the flooded passages linking the two.

The success of this exciting link-up by divers was a result in no small part of the work of another group of dedicated

specialists: the cave-surveyors. If the top and bottom ends of a cave system consisted of single stretches of dead-straight stream passage (an uncommon situation, it has to be admitted), it would be reasonable to look for the connection between the two past the known ends of those passages. But in a complex system, such as that which Gaping Ghyll and Ingleborough Cave comprise, there is rarely such certainty.

To have any chance of making sensible conjectures as to where effort should be concentrated – which sumps, which climbs, which boulder chokes – cavers must have accurate surveys. True, major discoveries are fairly often made in quite unpromising and 'illogical' places, but less so when it comes to uniting two known caves.

In the mid-1970s, the Kendal Caving Club started a major re-survey of Gaping Ghyll. Their painstaking work revealed inaccuracies in previous, rather hotch-potch, compilation surveys. To speed up the process of finding out exactly how the ends of the two caves lay in relation to each other, it was decided to pin-point the two ends by radio-location and then survey around these points to a high degree of accuracy. The equipment used was the Molephone, a brilliant piece of apparatus invented by Bob Mackin, a technician at Lancaster University. Not only does the Molephone permit a very accurate fix to be made on the surface by a transmitter underground, it also offers a direct two-way speech link – very useful in co-ordinating complicated diving efforts in two different directions, and in transmitting flood warnings.

In 1982, with the help of the Molephone, surveying revealed just how close the two caves were to each other – much closer than imagined, in fact only 4 feet horizontally and 11 feet vertically! Inspired by this information, two separate teams descended the caves, one to the Spiral Aven Series in the far reaches of Gaping Ghyll, the other diving through to Radaghast's Revenge, in Ingleborough. The intervening boulder choke was attacked with vigour and eventually an outstretched hand from one side grasped an outstretched wellington boot thrust from the other.

Two more digging trips were necessary before a through-trip could be assured, and then, during the spring of 1983, the teams involved were delayed time and again while they awaited the

coinciding of favourable weather conditions and the availability of the specialist film crew that was to record the whole episode. On 28 May, the attempt went ahead. Geoff Yeadon and Geoff Crossley entered Gaping Ghyll as part of a 13-man team; Jim Abbot and Julian Griffiths waited for the signal, then walked to the long sumps of Ingleborough Cave, their diving equipment carried by a support team of four.

Abbot and Griffiths finned their way

through the sumps, following the guideropes fixed by succeeding generations of divers who, over the years, had pushed the flooded unknown a little further back. At Radaghast's Revenge, while the film camera whirred, they exchanged diving equipment with Yeadon and Crossley. The latter pair dived through the sumps and were escorted in triumph out of Ingleborough Cave by members of the Bradford Pothole Club to the sudden

Above A small community of tents springs up at the great maw of Gaping Ghyll. The winch gantry can be seen at the bottom of the picture.

blinding lights of television crews and a champagne reception. It was four and a half hours later before the last man on the other party was winched to the top of Gaping Ghyll's main shaft — exhausted, but, like every caver involved, elated. The promise of Gaping Ghyll was realized.

41

Juniper Gulf

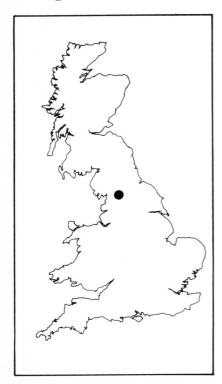

The windswept fells of Ingleborough conceal with their tufted folds the entrances to many classic potholes. Some, like Gaping Ghyll, declare themselves boldly. Others, such as Juniper Gulf, open suddenly beneath the feet, invisible only a few yards away.

Juniper Gulf commences as a broad sudden slash, black and gritty, a landscape wound that drops sheer for 90 feet to a rubble floor well sprayed by falling water. It is bright with daylight but this quickly dies into gloom as the water is followed into the hillside through a slim sinuous rift, the floor descending briskly in a series of short falls until the way becomes impossibly narrow at floor level and the explorer must resort to traversing along ledges and bulges on the walls.

The caver is now in a half-way world, roof and floor invisible, progress limited to straddling black space that occasionally yawns wide enough to make him tingle with adrenalin as he carefully shuffles across resounding gaps, supported solely by friction against dry grey walls the texture of starched hessian.

After a while the passage begins to close in, or rather the caver finds himself up near the roof, and a slot between the leaning walls gives access to a spacious pitch some 90 feet deep that

Above A caver abseils from daylight into Juniper Gulf.

swiftly opens out into a superb Yorkshire shaft, roughly cylindrical with smooth fluted walls the colour of golden syrup, and glistening with spray from the stream hurtling downwards. On ladder or single rope the caver passes straight down the centre to land on a level floor in a massive passage littered with vast chunks of limestone, natural black or brown masonry partially demolished and washed clean.

Scrambling through the boulders to regain the stream which has cut through the depths of the rubble, the explorer is suddenly confronted by an intimidating vertical drop. Where water spills over the edge a descent is only feasible in settled weather, but by re-ascending the

boulder heap, access may be gained to a gravelly balcony where a careful climb leads to a magnificent free-fall of 200 feet down the most beautiful shaft in Britain.

Narrow at the top, the pitch walls rapidly spread apart as the climber enters a lonely world of space distantly confined by huge soaring flutes in the pallid walls and the tiny pin pricks of light dancing far below him as comrades walk the floor of the awesome cavern.

Eventually the floor is reached and the stinging falling water leaping eagerly into a clean rift passage may be followed for a final 200 feet down a succession of short steps until the way is filled by a clear green pool, 420 feet from daylight, where the roof dips underwater and men can go no further.

Lancaster-Easegill System

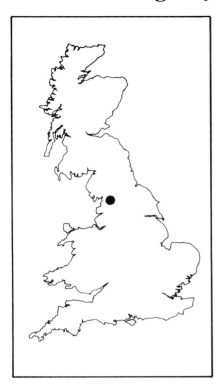

No reference to Yorkshire caves would be complete without mention of the phenomenon that has come to be known as the Three Counties System because its winding ramifications underlie parts of Cumbria, Lancashire and North Yorkshire. Discovered almost by chance in 1946, it has been an irresistible attraction for exploration ever since, offering every sort of caving challenge – climbing, crawling, diving, huge caverns, tiny passages, might boulder piles and deep shafts. It is a speleological paradise.

The considerable area of Lancaster Hole is, quite simply, one of the great wildernesses of Britain. A main trunk passage, huge beyond belief in parts, lies beneath Cumbria's Casterton Fell, and contains many indications of the last ice age. The legacy of those violent times is evident in colossal hillocks of fractured rock, in rolling waves of compacted mud, and a jumbled wedge of rubble 100 feet above the present-day stream, which burbles beneath glowering caverns in a splendid cockled waterway, oblivious to the devastation that forms its roof.

The hugeness continues when experience from previous caves suggests that it must stop. Different sections, revealed by careful exploration, display a multiplicity of characteristics. Here a vast expanse of soft loam and stumpy stalagmites, like some surreal war cemetery; there an enormous slope of boulders leading down to a river passage far below. Not just passages but whole systems lead off the main route, meandering out with dendritic complexity to touch, finger to finger, neighbouring caves and potholes.

Not far over the moor from Lancaster Hole the prominent streambed of Easegill cuts a wavering incision through the heather. Most of the time it is bone dry, and only a wasteland of sagging limestone cracks betrays the existence of caves beneath. All the entrances here have been cleared by excavation, in a series of digs dating back to 1946–47, when the first breakthrough was achieved in Oxford Pot.

Below The unparalleled splendour of Easter Grotto.

Lancaster-Easegill System

Discoveries came thick and fast. The Easegill Caverns comprise a fantastic collection of zig-zagging scalloped fissure passages, some wet, some dry, and varying in colour from fawn to chocolate. A myriad different ways snake west from the gill, coalescing in the main Lancaster-Easegill river passage. Opportunities for variation are endless. Cavers have the choice of climbing jagged rents in the roof to old wide caverns festooned with stalactites, or scrambling up masses of steep rubble into the main Lancaster system and onwards, or penetrating to deserted chambers and crawls far upstream, hours away from the well-worn routes.

Many miles had been mapped, but cavers are never satisfied. There was one burning question: could the caves continue *under* the stream valley to link with the fabulous systems on nearby Leck Fell?

In the late 1960s, a great discovery was made down one of these adjacent systems, Pippikin Pot, after a dreary succession of blasting operations and the forcing of very tight fissures. The reward was a sprawling network of old tunnels reaching ever closer to Easegill beck. Soon after, cavers in Lancaster Hole pushed through muddy crawls to enter a complex of large caverns ending at a group of deep pitches. Below, messy hands-and-knees passages ran on *under* the valley floor. Now work began in earnest, and in 1979 the hoped-for was achieved. A slot near the stream-bed was cleared, 60 feet of ladder fed down and climbed and Link Pot had been found.

It is a confusing place, but cavers soon passed down further shafts to marry the cave with Lancaster, and an incredibly muddy sewer in the opposite direction ended in a far-flung corner of Pippikin Pot. The major connections had been made. The long postulated Three Counties System was a reality, and, with 30 miles of passage currently surveyed, the longest cave in Britain (and the 14th longest in the world table) had been pieced together.

A through-trip, perhaps by exchanging parties, is now a much sought-after experience. If the chosen way down is through Pippikin Pot, the party, fresh and keen, will force themselves through very tight tubes and rifts with eagerness. Nothing dampens their enthusiasm for the well-decorated chambers and railway-size tunnels leading into the hillside – except, perhaps, a nagging knowledge of the long, long trek to the other end.

After a heart-pounding exercise in low crawls and boulder scrambles, another shaft – clean and echoing as conversation wakes the still chamber – beckons them down, ushers them on. A long hands-and-knees or stooping, back-breaking lurch follows, and then more shafts, dark and watery, hung with silver-wired ladders diminishing above the climber into a single soft line.

Once Lancaster Hole is entered, tiredness begins to bite. Weary cavers trudge the mottled floors of pounded mud, paying scant attention to the miracles of underground nature all around – great sweeping expanses of naked rock, broken by an outburst of stalactites, glowing in the lamplight.

Even the famous sights are only half-witnessed, such as the Colonnades, where stalagmites, like stone conifer trunks, prop up a ceiling traced in crystal, twinkling like snow. Daylight calls. The surface vegetation smells of a more familiar world.

One by one the cavers step up the last dull alloy ladder rungs, and – reassuringly tugged by vigilant lifeliners – begin the 100-foot climb, step by swaying step. At last, nudged out of their somnambulance by nodules of limestone, they struggle through a concrete manhole and emerge into the great openness of the surface.

A cave system of already great complexity, yet even now more discoveries are revealing potential for yet another great Yorkshire fell – Kingsdale – to be added to the sum. And beyond even this distant point, explorers are far from certain that the end of the road has been reached. Forty years on, new territory is still being found. The next forty could bring about finds just as extraordinary.

Below The great columns of Lancaster Hole, one of the best known groups of formations in the North of England.

Meregill Hole

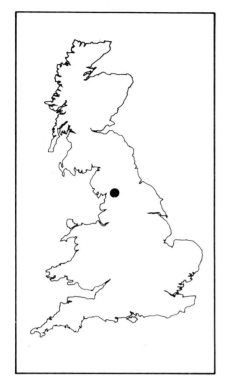

Some pots give up their secrets without a struggle, even when the entrance is small and obscure. Others, although open and obvious, yield after the manner of a knight engaged in trial by combat, reluctantly, slowly. Meregill is one such.

Meregill Hole, lying beside one of those ubiquitous drystone walls that straggle up Ingleborough, is a pleasant open shaft decorated with growths of stunted trees and sporting a deep sinister lake – the Mere – at the bottom. Except in very dry weather this mere stays constant at 40 feet deep, and the main entrance to Meregill is usually just under water, necessitating a free dive of 15 feet for the hardy caver who insists on it. Nowadays there are other entrances that bypass the sump.

The pot does not offer any of the usual caving trip hazards – tight squeezes, crawls, boulder piles and so forth – and a trip to the bottom would be remarkably free of effort save for the ever-present water. Meregill is a wild cave in the best sense of the word, wild and tempestuous. There is little danger of drowning, but on occasion a very real risk of being washed off ladders by the sheer force of the water.

The entrance passage rapidly gains in size through sculpted rock and ends at the first of three mighty waterfalls – 70 feet, 90 feet and 100 feet. Each of these shafts has a helical configuration, and is liberally provided with ledges to break up the climb, but all are lashed and hammered by numbing sharp masses of water and spray. They are not places to linger.

Below the second, 90 foot, fall, an impressive feature named the Canyon is entered; a huge rift 70 feet high at the start, noble and spacious, that ends above the superb 100 feet shaft whose noisy lower section falls sheer and unbroken to provide a really splendid ladder climb, lightly sprayed and just the right distance off the rock face.

Below, the cave irresistibly beckons the caver on with delightful irregular passages and short drops, and a rather deep pool where the unwary will find themselves undergoing an involuntary baptism by immersion.

With appropriate solemnity the cave enters the final phase in its 565 feet deep journey with darker coloured rock and a narrower, lower passage. With seemingly interminable twists and turns the cave descends almost imperceptibly, the only indicator being the stream rushing past one's feet. Suddenly, from a hole

Above Abseiling into the depths of Meregill Hole.

eight feet up the left-hand wall, a firehose torrent spurts across the passage with great force. Another major cave must exist nearby for such a volume of water to appear there, but exploration up the side passage revealed only a long series of sharp tight crawls ending after 2000 feet in a miserable silted stump.

Not until 1967 was the answer found when cavers pushed into an obscure stream sink not far from Meregill to be led, after an enjoyable series of short pitches, to a stupendous chasm swallowing water with a steady growl. This, the Black Rift, was duly bottomed – it is 270 feet deep – and after more sporting passages and climbs a sump was found. It has been dye-tested and proved to supply the Torrent Passage in Meregill. The new cave, Black Shiver Pot, lies on higher ground and a physical conneciton would make the combined system over 620 feet deep.

Penyghent Pot

High on the flank of whale-backed Penyghent mountain lies the unpromising looking entrance to what is, in fact, a rugged pothole known throughout the caving world as a superb and taxing underground system. Three decades have passed since its discovery, yet its reputation remains intact. Indeed, for many, Penyghent Pot represents the ultimate in the quality of its challenge: a tangible expression of the hard cave cult, requiring total commitment.

Cavers go underground for a variety of reasons, but one thing is clear: they go down Penyghent knowing that they will be engaging the triple alliance of cold, water and depth. Even on the finest summer day, the seriousness of the undertaking is soon felt, usually within minutes of negotiating the entrance.

This entrance is down-beat — an irregular climb in a crooked drystone shaft to a crouching chamber. At the back, the way is through a low black canal, busy digesting a brown foam-flecked body of water that stretches for over 800 miserable knee-wrecking feet before discharging itself down the first of a dozen well-sprayed shafts. And then the team is well and truly committed, for the canal is low enough to fill completely, especially when a freak summer thunderstorm rolls over the Dales.

Below, the pot continues rock black. The stream plunges into a soaring hostile chamber, whipping the air into a constant soaking swirl. And the pot has chosen this forsaken spot for its longest pitch — 130 feet, but broken by a shelf where the resting caver feels himself chilling by the minute, despite the wetsuit. It is the first place that the quiet warning bell sounds at the back of the mind: this is a place of exposure.

From here the stream rushes off down a sharply tapering rift, straight as a Roman road, and dropping relentlessly. Much can be free-climbed, but the wise ladder the hardest parts, remembering the exhaustion that will come on the exit.

Eventually a junction is reached where a silent menacing river slides out to be swelled by the boisterous Penyghent stream. For it is the latter that is the youth, almost accidentally cutting into the much older Hunt Pot drainage tunnel, sharing the final corrugated passage. The blackness of the walls and water, and the strong smooth push of the river against the legs of each caver contrasting with the earlier spray and clatter, all combine to give an overwhelming sense of depth. After two more short but spectacularly wet pitches, the pot dives to secret regions at a sump. Here you are 525 feet deep, some 1½ miles from daylight. It feels much more.

The return is all uphill. Punishing ladders have to be climbed, streams pummelling against tired legs, and eventually, when the last coil of tackle has been gathered up, there is the canal — black and sinister, cold and cruel. To exit from Penyghent is to experience total relief.

Yet the cave promises more — much more. For miles all around, the captured waters of the Dales, cascading down dozens of pots, unite deep in the stone fastnesses. There is probably more master cave sleeping below Penyghent Pot than cavers dream of.

Left Some of Penyghent's shorter pitches can be climbed with care.

Providence Pot to Dow Cave

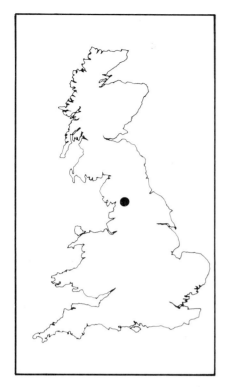

Notoriety has a way of attracting cavers – at least those with a taste for challenge at the keenest level – and this through-trip, while by no means the longest in the country, has had a reputation second to none since it was opened up in the early 1950s.

It was then that a partnership of determined explorers found a wet fissure disappearing from the well-trodden extremities of Wharfedale's Dow Cave until all that could be seen was a very high, narrow, glass-smooth rift stretching off into dark grey. Standing at this spot today, one realizes that they must have felt like mice encountering a door very slightly ajar. Once the decision was made, the battle was on.

Many weary trips and months later, great stretches of the dreadful rift lay mapped behind them. The way, though dead straight on paper, had not been easy. To navigate a possible route requires energetic forward shuffles, achieved by pressing opposing limbs against the two walls. Below, the slot pinches together, and woe betide the hapless victim who, running out of steam, slides down to wedge there, unable to free himself without considering help from team-mates or, in extreme cases, rescuers.

The explorers found that the rift, baptised Dowbergill Passage, con-tinued way beyond their expectations. Until, that is, it became clear where its water was coming from: the beck that tumbled noisily into a choked sinkhole in the next valley, 1 mile or so from Dow Cave. The cavers dug out the obstructions that choked the sink. Inside, a depressing series of mud-floored boulder cavities began at last to drop, like a continuous hole through a stack of black cardboard boxes. The day running water was found gave cause for celebration – the 4500-foot Dowbergill Passage was complete.

Above Dow Cave presents a vertical maze to the caver.

Today, the through-trip from Providence Pot to Dow Cave has been reduced to a mere two hours – but only by caving's super 'rock athletes', for Dowbergill demands the techniques of the high-tech gymnasium in abundance. It is no place for the beginner, the unfit, or those unsure of the complex and vexing multi-level route.

Giants Hole

With its wide range of passages, from active streamways to low muddy crawls, Giants Hole ranks as one of the most popular caves in Britain. Except during May, when the sheep grazing in the surrounding pastures are left undisturbed for the serious business of lambing, every weekend sees parties descending, from novice to hard-man – and mid-week too, for this is a popular venue for various outdoor pursuits centres.

The entrance to this classic swallet cave lies in the side of an open-ended gulley bounded on each side by low limestone bluffs. It swallows a small stream which gurgles down off the impervious gritstone of Rushup Edge – the twelfth such stream to disappear from daylight, hence the cave's other designation, P12.

We have no records to tell us who first poked a tentative nose into the large entrance, but the entrance series was probably fully explored before the Second World War. After, cavers took up the challenge in earnest, and even managed to pass what must have been viewed, in those days before wetsuits, as a distinctly nasty barrier: the first Curtain. Here the rock dropped to within 9 inches of the water surface, leaving the caver no choice other than virtually complete immersion.

A water obstacle which served as an even grimmer deterrent to explorers was Backwash Pool, an uninviting sump. This particular lock was finally picked in an ingenious manner by a determined band who constructed a series of concrete dams. The water from the sump was baled into the dams, giving access to the heart of the giant: and subsequent parties could use the same key again, although it was a very tedious and time-consuming business. Nowadays the process of cave growth has been accelerated by man. Backwash Pool has had the plug pulled out of it, and extensive blasting (in a now abandoned attempt by the landowner to turn Giants into a show cave) has destroyed the Curtain. Its teeth drawn, the entrance series is now merely an undemanding amble.

What was once the elbow-punishing flat-out Pillar Crawl is now an easy walking-size passage leading to Base Camp Chamber – still bearing its incongruous name despite the now easy access – above which the caver can climb to view a section of decorated passage. Following the stream from the chamber, past the impressive Boss Aven boring into the rock above, the caver comes to the 15-foot Garlands Pot – a short drop by any standards, but memorable because its base marks the start of Derbyshire's most notorious stretch of passage: the Crabwalk.

While any caver is used to meeting the occasional short stretch that requires a few sideways steps to pass, the 2258 feet of Giants Crabwalk is a joke gone wrong by going on so long. Were it a straight rift, disappearing thinly into the distance, the prospect would probably prove too much for many. But as well as its three long, slow, overall curves, the Crabwalk has chosen to grow as a sinuous rift with a multitude of small zig-zags, each bend in prospect offering relief 'just around the corner', even when the sideways shuffling caver knows that 2000 then 200, then 50 feet have to be strenuously sidled before the last 8-foot stretch really *is* that. Three intermediate practical jokes keep one from the conviction that the world will be forever sideways: the tight grip of the Vice, and the two awkward scrambles of Razor Edge Cascade and Comic Act Cascade.

No prizes are ever offered for guessing why the more respectably proportioned Great Relief Passage – which now leads the baffled caver to the sump – was so called. Just before the sump, a

passage leads off to the Eating House, and the caver of the 1980s is usually as ready to take a break here as his predecessors.

In contrast with the single-mindedness of the Crabwalk, there is a choice of routes on from here. An awkward climb up leads to the impressive Maggin's Rift, which rises steeply over a bouldery floor. At its start a passage can be taken to the Upper Series, an interesting conglomeration of walking passages, climbs and a wet crawl, eventually joining the Crabwalk at roof level about half-way along. The 40-foot climb down to the streamway is difficult enough to make this no easy alternative to the full lateral nightmare of the Crabwalk. But the Series does provide one very important link. Poached Egg Passage (which takes its name from a number of appropriately shaped and coloured stalagmite stumps) and a soaking flat-out crawl over 700 feet long tie Giants Hole with nearby Oxlow Caverns, a vast complex of half mine, half natural caverns, to give a whole system just under 3 miles long and a total depth – measured from the highest entrance, Maskhill Mine – of 793 feet.

From the Eating House again, the ultra-keen can bale the chilly contents of St Valentine's Sump to pass through to some remarkably muddy descents, aptly named the Filthy Five Pitches, and reach the bottom-most sump at East Canal. More will opt for the alternative route via the Lower Syphon Complex, a series of interconnecting hands-and-knees crawls leading back to the main stream again. A roof-level traverse provides an airy way into a passage to the Plughole – an awkward if short descent – and the 40-foot ladder pitch in Geology Pot to rejoin the stream. The dampness of the 15-foot deep Cascade is followed by an echo of the now-vanished duck in the distant entrance series in the Far Curtain. Even if its earlier cousin is now consigned to limestone dust, this curtain metes its revenge by forcing the visitor to duck through icy water. All of this leads to the hard-earned East Canal, plumbed to a depth of 100 feet, into which the stream is swallowed for an undisturbed and uncharted subterranean coursing before emerging to daylight at Russet Well, several miles away.

Right Climbing down into Base Camp Chamber.

DERBYSHIRE
Jackpot (P8)

Above A ribbed calcite flow in Jackpot.

A mile and a half south-west of the top of the steep plunge of Winnats Pass, on the bottom-most slopes of the much-quarried Eldon Hill, is the entrance to Derbyshire's most visited wild cave. Its full name, Jackpot, is nearly always listed first in guidebooks, yet cavers perversely persist in talking about P8, holding more affection for its original designation when it was a mere 'promising site'.

Like Swildon's Hole on Mendip (see page 60–61), P8 is popular with experienced cavers and novices alike and has witnessed the introduction of many a local caver to the sport of caving. Like its southern counterpart, too, it tolerates novices only when its stream is at normal level; in times of high water it becomes a far more serious proposition.

The stream which feeds the swallet entrance runs, like that in nearby Giants Hole, off the unyielding gritstones of Rushup Edge. Originally dug into 20 years ago, P8 offers ½ mile of passages entertaining enough for even the most jaded caver, and reaches a depth of 170 feet. In addition, beyond the Main Stream Sump members of the Cave Diving Group have added another 2000 feet of length, but only by meeting the tough challenge of, to date, eight separate sumps.

The sport starts from the word go with the climb down into Cascade Chamber, where you are guaranteed a liberal soaking. A fine stream passage follows, the water leading down a series of wet slides, to the brink of Idiot's Leap – an 8-foot scramble where the beginner often welcomes the offer of a handline. Now the streamway narrows to a wriggling rift, with more water shutes and pools thrown in for good measure, ending at the First Pitch. The degree of sport for the caver on this always wet 25-foot ladder climb depends entirely on the water level, varying from a thorough but entertaining soaking to a roaring torrent, dangerous for even the experienced, in flood conditions.

From the spray-filled chamber at the foot of this pitch there is a choice of two ways on. The stream can be followed again as it chatters along a narrow wet passage leading to the 20-foot drop of the Second Pitch, falling into an impressive boulder-strewn chamber. But the most popular route is to climb up a distinctly greasy slab above the stream to the Upper Series and traverse over gaping holes dropping back to the water. In an area quite well decorated with the timeless soft contours of flowstone covering, an iron ladder provides an easy descent to the boulder-floored chamber, where the alternative route is rejoined. Here, the entrance stream vanishes down a number of impenetrable rifts, its contribution to the cavers' sport over. It emerges at the surface at Russet Well, in Castleton, over 2 miles away.

Another, even larger stream beckons, but first there is a short sideways shuffle into the all too aptly named Mud Hall, before an easy scramble down to the water. Upstream, progress soon halts at the deep clear pool of the Upstream Sump. Downstream finds the passage gaining height, becoming an impressive river passage – but only to be followed by the non-diver for barely 300 feet, for then the roof dips relentlessly into a large dirty lake, marking the start of the 105-foot long Main Stream Sump. Plenty of other crawls and passages can be taken in on the way back to the surface, though, to round off the visit to this compact but entirely sporting and satisfying system.

Nettle Pot

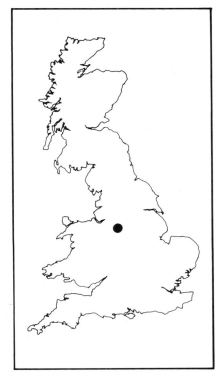

Take a random selection of meets lists from caving clubs in, or visiting, Derbyshire, and you will find Nettle Pot high in the popularity stakes. This classic pothole has very little horizontal passage; its effort has gone into piercing the Peak District limestone as urgently and directly as possible. As a result, a plan of it shows a virtually straight line hardly 800 feet long – but the cross-section tells the true worth of the system, with the shafts biting straight and deep, and the few passages striking off left and right halfway down almost as afterthoughts.

The pot's entrance is covered by an iron lid, not to hold keen cavers at bay but to prevent sheep from meeting a sudden and savage slaughter at the end of a sheer 160-foot plummet.

The first 60 feet were painstakingly excavated by a local caving club in 1934, and one can only assume that some judicious tying-on had been arranged by the brave person who made the final breakthrough! This initial section is extraordinarily constricted ('impossible to negotiate with a stretcher' warns the guidebook sombrely), and the slight opening up at the Sentry Box provides a chance for cavers to re-fasten the ladder more conveniently. An intermediate lifeliner usually takes up cramped position here, too, as shouted or

whistled instructions from the bottom are apt to be muffled to the point of extinction for anyone on the surface.

At the bottom of the entrance pitch, which widens into the more hospitable dimensions of the Bottle in its lower part, are the Flats – a washed-out lava bed, so wide in places that the roof seems to be impertinently defying gravity, and the caver silently prays that its defiance will hold good at least for the next few hours!

The caver now climbs down a small pitch and works his way along a hair-raising traverse across the top of his next objective, Elizabeth Shaft, to a boulder-floored chamber. A small hole in the floor of this chamber gives him the best 'hang' for the next breath-

Above Deep in Nettle Pot, a small hole leads to the top of the spectacular Elizabeth Shaft.

taking descent. The longest underground pitch in Derbyshire, the free 170 feet of Elizabeth Shaft provide a magnificent descent for the caver.

From the bottom, a mass of shattered boulders which have peeled off the huge heights above over the centuries, two short pitches lead to an intimidatingly tight slot – the Sting. This is as far and as deep as most go, but the thinnest and most determined can squeeze through to reach the base of the adjacent Beza Pot, the alternative route to these regions over 500 feet underground.

51

Agen Allwedd

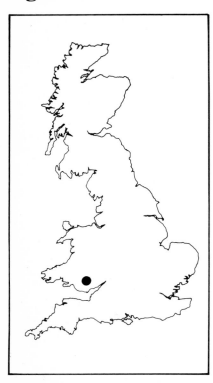

Frowning across a broad valley towards the busy village of Crickhowell is a thick dark crayon daub a short way below the round-capped tops of Llangattock Mountain. This is the dark sheer face of the long-disused quarry that left its all too visible mark on the mountain's limestone. Running along the base is the old tramroad, now much overgrown, and at its western limit casual strollers might be surprised to find a metal door set into the cliff, small enough to be the portal for some Celtic version of *Alice in Wonderland*. Surprise would be all the greater if they were told that beyond it lie 16 miles of sprawling cave passages and chambers.

Aggy, as it is affectionately known, draws two contrasting reactions from cavers. There are those, it has to be admitted, who find its long stretches of unchanging terrain less than exciting. But there are many more who relish its particular character: a place of ancient halls and long passageways demanding much perseverance of the caver.

The entrance passage behind that locked door gives no real clue to the nature of the cave. After signing the visitors' book and noting down your estimated time of return, you launch upon 1300 feet of generally walking-size passage with enough interruptions in the way of fallen boulders, stoops and occasional crawls to break any real attempt at rhythm. A small stream adds its voice towards the end, but hardly dramatically. Launching into the First Boulder Choke demands more cautious progress as you crawl or squeeze through the bottom-most layer of jumbled boulders, piled high around and above you.

The scramble clear of the choke brings you, with a shock of surprise, into the blackness of a great void — Agen Allwedd's Main Passage. This vast, bone-dry concourse runs for nearly 3000 feet along the edge of the mountain, and its great size invariably reduces cavers' banter to whispers. Two or three double-decker buses could be dropped in here side by side, with another couple on top — and there would still be room to spare. From this strong spine run the two major passages

(together with other smaller ones) that lead to the mountain's deepest secrets: Main Stream Passage and the notorious Southern Stream Passage.

The first brings a welcome relief from the dead, dry mud floor of the Main Passage, for this is a busy streamway, deceptively low at first, but soon opening up to easy strolling size. The only time-consuming obstruction as you wade into the heart of Llangattock is the Second Boulder Choke and its attendant high-level traverse, where a fixed steel wire provides a reassuring safeguard against a slide into the black slot dropping sheer to the distant streamway. The waterway is rejoined and followed to Northwest Junction, an important parting of the ways. Here the moderate stream followed so far joins the more earnest flow running from right to left.

Upstream, the fine waterway can be waded for well over 1 mile until the walls pinch in and the roof dips into Turkey Sump. From here the cave-divers must take over, but well before this demarcation-point two passages lead off from Turkey Passage, each giving access to the dry rambling ramifications of the Summertime Series, with its sandy caverns and pockets of glinting selenite crystals.

Downstream from Northwest Junction, the going becomes much more serious, for here there is no respite from the water and the distances in terms of caving hours mount mercilessly. Nearly 3000 feet of stream passage must now be negotiated before the Third Boulder Choke is reached, and there is a sting in the tail at Deep Water and The Narrows. In the 1960s, cavers used inflatable dinghies or even inner tubes of tyres to pass these sections, but today's caver, enjoying the extra buoyancy and warmth of a wetsuit, simply wades and swims through the deep, slow waters.

As recently as 1972, the Third Boulder Choke marked the end of all trips downstream in this direction. But in that year the Eldon Pothole Club from Derbyshire put the finishing touches to years of excavation by the

Left Rare mud formations in the Main Passage are guarded by tapes.

Above Stalactites in Turkey Streamway.

Chelsea Speleological Society, and found the long-sought link through to the lower section of streamway, passing a fourth and fifth choke in the process. Few will be able to resist following this superb final streamway for its full 3000 feet over several small but refreshing falls before the stream disappears into the maws of a long sump. After a brief rest at this enigmatic place, the caver can, of course, simply retrace his route in, but many will brace themselves for the most demanding part of the 'Grand Circle' tour – the exit back to the Main Passage through Southern Stream Passage.

On the cave survey, Southern Stream Passage shows as a zig-zagging scratch line; in the rock it presents itself as a damp, malevolent and never-ending passage, an uncomfortably close if not tight fit on the caver. Every few yards comes a different obstacle – a stoop here; a sharp bend to the right there, now to the left; a hands-and-knees crawl; a tight rift, requiring crab-like motion; and innumerable stones to clamber over. None of these is exceptionally demanding technically, but strung out over the passage's grotesquely long 6000 feet they add up to a bleak and demanding journey with only one briefly entertaining interlude in the form of a small waterfall climb.

After years of relatively few new finds, Aggy cavers are now enjoying a renaissance as the wakened giant grudgingly gives up new passages and chambers. The potential for more, much more, is great.

Ogof Craig-a-Ffynnon

The clues to the existence of this superb cave were always tantalisingly present. For years cavers scrambling along the Clydach Gorge in Gwent were fascinated by the immense flow of ice-cold water which gushed up after heavy rain in a whole series of springs over a ½ mile stretch of the gorge near the Rock and Fountain Inn. The temperature told much: for the water to be that chilled, it must have travelled a good distance underground, losing its heat to the great mass of limestone above.

The prospects for cavers were obvious enough, and tempting enough, for a half-year digging operation to be launched in the late 1950s. But this determined onslaught could not crack the secret of the cold springs. As is often the case with large cave digs, eventually talk in the pub turns to the possibility that just that *extra* bit of effort, or digging in a *slightly* different place, might be all that is required. The attack on the hillside was started again in a disused quarry in 1973. It says much for the faith of cavers that the same small team kept at their thankless task for year after year, their eventual reward only coming in 1977. But what a reward!

In the intensive explorations that have continued unabated since that breakthrough, Craig-a-Ffynnon (which takes its name from the nearby hostelry)

has revealed nearly 8 miles of fine mature passages and huge chambers harbouring some of the most spectacular stalactites and helictites in Britain. This is a system which reveals dramatically the three-dimensional nature of caves, for in the section so far known it has cut through the stone on two distinctive levels. The higher passages are those that the stream formed first. These were abandoned by the stream in the later stage when the lower passages (still active today) were first created. (This common process is described on pages 10–12.)

Worthy of the best caves anywhere in the magnificent Hall of the Mountain Kings – 95 feet wide and up to 70 feet high, and decorated with calcite on a scale that properly merits the word lavish. Guarded by a 42-foot ladder pitch is the Lower Series with its helictites of a foot or more in length – a far cry from the more usual couple of inches reached by these gravity-defying formations. For the caver who sees beauty in rock shaped by water, the sculpted profile of the sandy-floored Severn Tunnel is unforgettable.

Such beauty is, for the caver, worth all the effort and risk. Risk there is, too, in this spectacular place, especially after dark rain clouds have settled over the broad mountain tops, for then the streams can burgeon with alarming speed, flooding the entrance passages to the roof.

As exciting as the last few years have been in the unfolding of Ogof Craig-a-Ffynnon, the future promises even more, for now there is another clue – the draught which flows enticingly and healthily along North-West Inlet Passage. Cavers need no reminding that that way, further into the great block of Llangattock Mountain, lies the huge cave of Agen Allwedd (see pages 52–53). The dream is the forging of a link between the two – and the realization of such dreams is what, after all, drives every explorer.

Right Stalactites in Ogof Craig-a-Ffynnon. The scale in this large chamber is deceptive: the formations are over 20ft long.

Ogof Ffynnon Ddu

Water – in the unexpected place or in unexpected quantities – is one of the greatest hazards British cavers face. Yet in any trip it is also the added ingredient that adds the strongest element of immediate excitement, particularly when it presents itself in the form of a vigorous streamway. For the *afficionado* of underground waterways, Ogof Ffynnon Ddu (Cave of the Black Spring) is nothing short of Mecca.

As the crow flies, it is less than 2 miles from Pwll Byfre, the sinkhole that greedily swallows all the available surface water, to the Black Spring itself, where the cold water spills out in the upper reaches of the Swansea Valley. Yet between those two points it has, over thousands of years, sought the yielding weaknesses of the limestone to carve out a great complex mapped for 25 miles, the second longest in Britain, twentieth in the world. At just over 1000 feet, it is also Britain's deepest.

Considering its current status among the great caves of Britain, exploration of Ogof Ffynnon Ddu has been relatively recent compared with most of the major Yorkshire systems. It was only in 1946 that it began to command the attention of the then newly-formed South Wales Caving Club. Members marked their interest clearly by establishing their headquarters in a nearby row of old cottages. Such intimate proximity undoubtedly encouraged the unravelling of the system's secrets, from a cave hardly more than a couple of hundred feet long then to today's giant. But being on the very threshold did not make OFD yield overnight; it has been a near-epic tale of years of digging, diving, climbing – and painstaking surveying, translating the true state of affairs underground on to paper. Only then did cavers see the potential of those unpushed crawls and boulder chokes that could, and did, unlock another part of the puzzle.

Because the first known entrance was at the bottom, near the spring, exploration was a topsy-turvy affair, working steadily upwards. Low miserable crawls might have deterred even the keenest caver, but this entrance was found to lead directly to the superb main streamway.

This stream passage is the kind that all cavers dream of discovering: wide enough to maintain the water level between knee- and waist-depth, and in places soaring to a ceiling beyond sight. The walls, scooped by the bite of the torrent, are a delicious glistening black, calcite intrusions here and there providing lightning highlights of pure white. For added zest, in the floor of the streamway lurk four large potholes, concealed by the water as snares are hidden by a gamekeeper, ready for the unwary.

In times of flood, the clamour is intoxicating, each splash and cresting wave echoing a thousandfold. Yet this is no place for heroics at such times – the stream has already laid claim to life – and caving parties bless the network of high and dry passages which provide escape routes to the surface.

Two more entrances were discovered

as the explorers pushed further and further up the deep heights of Fforest Ffawr. Each provided the key to further discoveries.

At first the streamway led for only some 1300 feet before a boulder collapse and sump stopped progress. It first fell to divers to enjoy an extra 1¼ miles of the main stream passage in the early 1960s. Ordinary cavers were frustratedly envious at the reports that this section was even better – there were several cascades to be clambered up, a potholed stream bed, and the exhilarating Marble Showers, a place where water sprayed down from high-level passages, lashing the white-streaked jet-black limestone. Then, as things go underground, came the inevitable stop at a sump perched at the top of a 23-foot high waterfall.

Today, though, the caver can reach these further parts by wriggling through the sandy-floored crawls of Cwm Dwr Quarry Cave – OFD's second entrance – completely by-passing that bottom sump.

From the streamway, the original explorers looked high above them. The next revelation had to lie that way. Scaling poles (maypoles) were erected time and again, then the right place was found and Maypole Inlet proved to be the doorway to a baffling three-dimensional maze of passage, halls and canyons.

A trip through here reveals a feast of excellent formations. Apart from arrays of stalactites and 'mites, interspersed with twirling helictites, there are beautiful crystal pools, their water long since evaporated leaving priceless linings of calcite gems. And in a cul-de-sac, mirrored by a still, shallow pool, are the Columns, the most famous group of formations in Britain. Over a

Above The Columns of Ogof Ffynnon Ddu – superb formations which are now safeguarded by restricting access to certain days in the year.

Left A waterfall cascades from the roof at Piccadilly in Ogof Ffynnon Ddu.

dozen man-high columns of pure white calcite link ceiling with floor, with slender straws dancing between, counterpointed by short stubby red-tipped stalagmites. This is a place of wonder.

With the finest single streamway in the country and formations of incredible beauty, Ogof Ffynnon Ddu has properly been described as a great treasure chest. Small wonder that in 1976 it was given the greatest protection possible against encroachment by quarrying by being declared a Nature Reserve – the first in Britain never touched by the light of day.

Little Neath River Cave

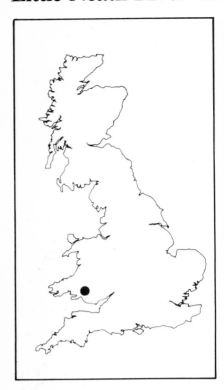

One glance at the entrance to this cave would be enough to confirm the worst fears of most non-cavers, and is sobering enough for the caver on his first visit. As you walk upstream a short distance from where the River Neath usually sinks among boulders, by Pont Nedd Fechan, it takes a reasonably keen eye to spot the tiny slit in the far bank, down which a copious quantity of the river slops even at normal water level.

This slit leads immediately to a flatout wriggle through fast-running water, with the added difficulty of a number of sharp bends to be negotiated. These difficulties rule out visits in anything but dry and settled weather, of course. But unexpected rain is not all that extraordinary in Wales, so once inside the cave cavers keep a constant and close eye on the water level, ready to exit rapidly, before the river rises those few extra inches and seals off the entrance.

Yet Little Neath, despite this formidable threshold, is a very popular cave. Rightly so, too, for its almost 5 miles of passages make up one of the most sporting and satisfying systems in Wales. The initial constrictions soon give way to a larger passage, generally large enough for walking, until, after some 1300 feet, a large sandy-floored chamber is reached at an abrupt T-junction. From the right flows the rest of the Little

Above Crawling in Streamway.

Right Canal Bypass, Little Neath River Cave.

Neath River, but a 60-foot sump passable only by divers stands between you and an easy way back to the surface through the adjacent Bridge Cave.

On the left the river, now whole again, spreads much broader as it courses into the wide and decidedly low portal of the Canal. Though this section can be by-passed by back-tracking some distance and following a small dry passage, the wet-suited caver always hopes for dry enough weather to allow entry to the Canal. With the roof only inches above his helmet, he can float the full 500-feet length stretched out flat, the current gently taking him ever deeper, steering by means of an occasional flick on the bottom with his fingertips.

The Canal and its by-pass meet at the sudden great space of Junction Chamber, 70 feet wide and 35 high. This heralds the start of almost 2000 feet of fine streamway of ever-increasing size, interspersed with large chambers. In one of these, where the

river is particularly invigorating as it dashes between rocks and over small falls, the name Bouncing Boulder Hall gives warning of the loose nature of the immediate terrain.

Tucked away at the top of a high bank of dry mud and silt flanking the streamway is the low crawl leading into the cave's treasure chest: Genesis Gallery, with its gorgeously decorated avens. This entrance is difficult to find, which, because of the fine formations it leads to, many cavers do not regret.

For the non-diver, Little Neath terminates where the stream passage dips to form Sump Two. Only the experienced cave-diver can pass this and the following two sumps to witness the majestic dimensions of New World Passage – up to 80 feet wide and 50 high.

Swildon's Hole

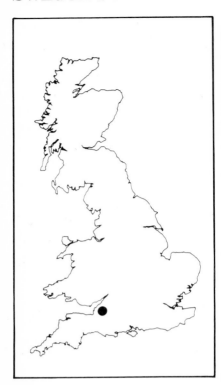

Most British cavers have a soft spot for Swildon's Hole, and anyone on Priddy village green on a Saturday or Sunday will testify to its popularity as party after party parks, clambers into caving gear in the traditional barn, then walks the short distance over several grass fields to the entrance. Here, at the foot of an unappealing stone, built to give some measure of temporary water-control in times of severe flood, the busy stream sinks into the honey-combed Mendip limestone.

Swildon's is not only historically significant as one of the first caves in Britain explored in systematic fashion, it is also a classic swallet cave system in its own right, providing a challenging journey down a long meandering streamway, almost 5000 feet long, and the passage by free-diving of several sumps.

While some caves can only be entered after months or even years of digging, Swildon's was always open. It was first entered in 1901 and the entrance series explored to the notorious Forty Foot Pot. This waterfall presented a difficult obstacle to the Mendip cavers of that

Right Following the stream along Wet Way.

time, ill equipped with candles and without the distinct advantage of wet-suits, and it was not passed until a drought in 1912. Until the late 1960s, the Forty still represented a psychological barrier, and many older cavers will remember the numerous 'mini rescues' occasioned by miscalculating cavers too tired to climb the wet pitch.

The installation of a short pipe to divert water from those climbing the ladder created a controversy that raged in caving journals until the Forty met its end. Just as a new generation of wet-suited cavers was about to debunk the myth of the Forty, Nature took a hand by washing it away in violent thunder-storms in 1968. The ancient passage under the pitch was excavated by the floodwaters to leave the Forty a shadow of itself, a mere 10-foot free-climbable waterfall. A sad side-effect of the loss of the Forty, it must be recorded, has been a noticeable increase in the despoilation of stalagmite formations deeper within the cave.

There is a choice of routes from the splashing entrance to the site of the Forty, varying in degrees of dampness. The Wet Way is probably the most popular, despite – or more probably because of – the antics involved in negotiating the convoluted and constantly flushing Lavatory Pan.

Beyond the Forty, an attractive 10-foot wide and lofty streamway leads to a shorter pitch, the Twenty. There are often queues of cavers at weekends waiting to pass this bottleneck, though it can be free-climbed by the more confident caver. Below the Twenty the cave continues down through cascades and deep pools including the famous Double Pots, both of which can be traversed over without any great difficulty, although at least one accidental plummet into them is held to be necessary before one is a true Son of Mendip!

Round the next right-angled bend the passage pinches in to a narrow inclined rift. This can be by-passed via the high-level Barnes' Loop, a once gorgeously decorated casket of formations. Though tarnished somewhat, it still offers a fine spectacle, and conservation-conscious cavers put up with the difficulties of the rift rather than risk accidentally inflicting any more damage. Only a little further another climb up from the chattering stream leads to another pocket of good formations in Trat's Temple.

Meanwhile the stream continues its descent, now along a large dark and gloomy gallery, 400 feet below the entrance, to Sump One. This is probably the best-known sump of all among British cavers, for it has introduced thousands to the breath-taking and -holding art of free-diving.

Sump One is a place steeped in caving history. It was the first British sump to be passed using cave-diving apparatus (albeit incredibly primitive by today's standards), and it saw the birth of the Cave Diving Group, still going strong 50 years later.

Although as forbidding-looking as any sump, the roof plunges below the surface of the dark water for only a few feet. Passing it is simply a matter of wading into the deep pool, taking a deep breath, and ducking through with one hand following the permanent guide rope secured on each side. Simple, yes, but frightening enough for many, and more than a few first-timers have had to be strongly 'coaxed' to brave it for the second time on the return journey.

Beyond Sump One the streamway continues for roughly ¾ mile, but its present extreme point, Sump Twelve, can only be reached by experienced cave-divers. Very experienced cavers can free-dive the first three sumps leading to this point, but these are longer and more serious undertakings than Sump One.

Also beyond Sump One is an assort-ment of high-level passages which are the remnants of the old stream routes before the water found its present level. Often extremely muddy and choked with silt, they have provided numerous opportunities for explorers to extend the cave by digging out the choked sections. In doing so, by-passes to the early sumps, of varying severity, have been discovered, enabling a number of round-trips to be made, including the now classic Round Trip, which involves passing several flooded digs by baling them and squirming through with one nostril in the air. These are extremely sporting (that is, taxing the caver's skills to the full), and in recent years cavers have often treated the Round Trip as a sort of time trial in high-speed caving – a good time being under two hours, which cavers of earlier generations would have deemed impossible.

Another inlet series offers the daunting prospect of a hair-raising traverse over a 40-foot deep hole to enter it. Its name, the Black Hole Series, says it all.

Even now, eight decades after it was originally entered, Swildon's Hole still offers the chance of new finds and – who can tell? – the sunless way through to Wookey Hole cave, where the waters finally come to light once more.

Below The quiet beauty of Tratman's Temple.

St Cuthbert's Swallet

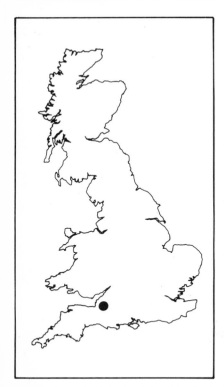

Not far from Swildon's Hole (described on pages 60–61) is St Cuthbert's Swallet, a system almost as long (well over 4 miles), and with a great richness of passages, chambers and formations. Yet St Cuthbert's, which takes its name from an old lead works nearby, enjoys its strongest reputation locally.

There are several reasons for this. First, access to the cave is strictly controlled by a Mendip caving club on behalf of the landowners so that only relatively experienced cavers in small parties, led by an approved leader, are allowed into the cave. Second, the cave has, so to speak, the sting in its head; unlike Swildon's, where easy caving precedes the more difficult, St Cuthbert's presents a high degree of difficulty in its earliest stages.

It is also extraordinarily difficult to describe any trip in St Cuthbert's to anyone who does not know it very well indeed. True, the principal guidebook, *Mendip Underground*, states simply:

The cave consists of a steeply descending entrance section which carries the main stream beneath a multi-level network of passages, chambers and boulder ruckles. At a depth of 350 feet, the system simplifies to become a single, gently sloping passage carrying the main stream, now augmented by several small inlets, to a final sump.

But this — paramount to saying 'London is a large city in England' — is followed by thousands of words of detailed route descriptions. The cave is fiendishly complicated, an intricate three-dimensional maze.

St Cuthbert's begins as a concrete barrel shaft sunk through boulders. A short descent leads to the intimidatingly tight 30-foot deep entrance rift, pinching in to as little as 1 foot wide, down which the stream flows. In very wet conditions this deep slit is impassable, and even in drier weather the water has to be diverted on the surface so that the rift can be passed comfortably on the return, when it represents the strenuous finale to the trip.

A short wriggle through boulders leads to a 25-foot ladder pitch beyond which the cave begins to demonstrate its complexity by branching into two

Above For this caver, the 'sky' is rank upon rank of graceful stalactite curtains.

Right Squeezing carefully past formations in Eastwater Cavern.

routes. The cave continues to descend in a series of pitches. Although the original explorers needed awkwardly large quantities of tackle to descend the cave, little is required on the main route now because fixed steel ladders have been installed. St Cuthbert's is one of the few caves on Mendip where certain fixed aids have been considered acceptable, but even so over the years many such aids deeper inside the system have been removed.

The steep entrance series halts at Mud Hall, a multi-level chamber from which a bewildering series of routes departs. There are so many different ways of reaching the same place in the cave that

Eastwater Cavern

it is possible to cover large areas of the system without retracing one's steps. Equally, one of the system's attractions is that average cavers can be exploring new (to them) territory very soon after entering the cave. A corollary to this, of course, is that it is easy to get totally lost very quickly.

St Cuthbert's offers a great variety of underground scenery, and there is something to delight every kind of caver. A route from Mud Hall to the Central Boulder Chamber provides the main jumping-off point for journeys up side passages. It is impossible to describe all the series. Choice examples are the exhilarating ascent of the Maypole Inlet Series (a number of climbs up short pitches, in a stream); the tortuous September Series leading to the beautiful September Chamber and its balcony of pure white stalagmites and stalactites; and the aptly named Railway Tunnel which ends in a 30-foot high white flowstone cascade. Boulder Chamber, one of the largest in the cave, has a boulder floor raking steeply down for 100 feet or so, and soaring up from its distant basement is the truly awe-inspiring formation of the Cascade. This flow of pure white calcite, fringed at the bottom by great organ pipe swellings, is a staggering 100 feet high.

Deeper inside the system all routes unite to enter a unique place – Gour Hall. Formed on a fault line, this lofty chamber is occupied by a stupendous 30-foot high gour – a beautifully curved and minutely pocketed natural calcite dam, a formation usually measured in Britain in inches but here so large that the caver must climb carefully down its glistening face as on any rock climb.

The passage continues straight as a die, following the fault, until it abruptly halts where the stream disappears under a wall. Until 1969 this was nearly the end of the cave, the water flowing almost immediately into the silted-up Sump One. However, over many years of digging, damming and ingenious shoe-string mechanics, the sump has been passed to St Cuthbert's II – 900 feet of lofty streamway. It ends in the tiny Sump Two where current digging activities are centred.

With more than 1 mile between the bottom of the cave and the nearest parts of Wookey Hole cave, where the stream emerges, together with that from Swildon's Hole, there is still much to discover.

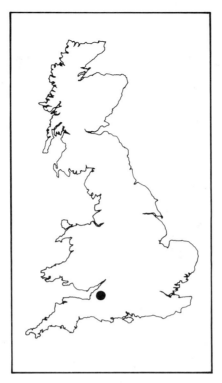

On Mendip this cave enjoys something of a sombre reputation. Because of the steep dip of the strata hereabouts, many Mendip caves work their way down quite soon, even quite precipitously, but the most extraordinary feature of Eastwater, the memory of it that may well linger most strongly, is the relentless *tilt* of the place, as though some god in ancient times became bored with the horizontal and jacked the whole thing up at one end until it inclined giddily at an angle approaching 45 degrees.

If any memory rivals this it will be that of the grim entrance to the cave. When first excavations began on the surface here as long ago as 1902, the diggers were appalled to find, instead of the anticipated open shaft or passage, an enormous jumble of huge boulders, piled willy-nilly in an unsettled malevolent buried heap. Even in that year the boulder choke – or ruckle, to give it its proper Mendip name – fired off a warning shot, temporarily trapping two of the diggers. Despite occasional attempts to stabilize it using explosives, it still, from time to time,

Eastwater Cavern

rumbles with discontent, with fatal results for one caver in 1960.

Such is the start, a climb of some 60 feet down through the choke,with a marker wire laid along the least unstable route, to the 380 Foot Way. A short way down this is the junction with the Upper Traverse − the key to the whole system. The Traverse runs along the top section of a very wide bedding-plane, low enough for the caver to find himself using holds in the roof to aid his slithering progression as much as those in the smooth floor. As well as the problem of getting himself from one side to the other without sliding down to jamb ignominiously in the tapering lower regions, he has the added problem of keeping a constant tight grip on the tackle bag. Equipment released here will invariably slither out of reach.

At the T-junction with the Canyon at the far end of the Traverse, the party can travel either upslope or downslope to reach the lowest parts of Eastwater, and there is a choice of two separate routes down to the Lower Series itself: via Dolphin Pot, involving a 40-foot ladder pitch, or the Twin Verticals route, combining free-climbing and two 30-foot ladder pitches.

Most of the Lower Series has been well-known for many years, but as recently as 1983 two cavers digging in the very unstable Ifold Series there discovered the 2000 feet of Westend Series, taking the total length of the cave to well over a mile, and its depth to about 460 feet. This discovery contains some fine formations, including the exquisite Regent Street − a 30-foot long, emerald green lake.

Back to that T-junction at the end of the Upper Traverse . . . Scrambling downslope from here, in a trench scoured in the floor by some long absent stream, takes the caver to Hallelujah Hole, a tongue-in-cheek name for a rather awkward wriggle through a tight constriction.

Below is the 150-foot-long, cramped Primrose Path which, following the relentless slope, ends in Tether Chamber. This 'chamber' (more cupboard sized) lies only 20 feet from the top of Primrose Pot. That 20 feet, however, is a low squeeze, tight enough to forbid entry to cavers of more than average bulk. With gravity and smooth rock to help, the descent of the squeeze is awkward enough; the return, with both working against the caver, is a very different matter, often requiring a friendly tug on a rope from above.

Falling sheer from the end of the squeeze is Primrose Pot which, at 185 feet, is easily the deepest on Mendip and ranks amongst the deepest in the country, though it is usefully broken into three sections. But a descent leads to only a few feet of passage, hard-won indeed, and meagre reward when the next move for the caver can only be the start of a slow return through the series of obstacles and difficulties piled up to guard this decidedly remote spot.

Below Manoeuvring a rope through one of the low sections.

Uamh an Claonaite

Far above the deserted Scottish glen of Allt nan Uamh near Inchnadamph, nestling in a broad sweep of weathered peat hags and stunted heather, lies the pretty lochan of an Claonaite. Although not exceptionally large by Highland standards, it has a unique and fascinating secret: many burns flow into it, but none flows out. The explanation is simple. Claonaite lochan is greedily swallowed up by the thirsty Cambrian limestones of north-west Sutherland to be discharged into the sun right at the foot of the glen at an astonishing river rebirth called the Fuarain Allt nan Uamh – Great Spring of the Burn of the Caves.

This promising site was denied to cavers until 1966, when a creaking tottering agglomeration of white rock was gingerly excavated and squirmed through. A roaring streamway was found, leading on into the hill.

There are two main types of limestone in the Assynt region, one whitish yellow, the other gun-metal grey, and both are peculiarly knobbly. Besides carrying a large and freezingly active stream, substantial sections of the system demand hands-and-knees work that can prove damaging to both clothing and flesh.

The streamway is acclaimed as providing one of the best underground journeys north of Yorkshire, and the sheer variety of obstacles provides deep satisfaction for the caver. There are chest-deep pools edged with foam, flat-out wriggles where the caver's nose is just above the water (usually), climbs down foaming cataracts, plus ramps where the stream fans out, roaring with enthusiasm as it dashes itself to the bottom. And there are sumps, six in all, but only two that demand diving. Sump 3 is particularly attractive. Here, a steep climb down wet, sandy boulders gives on to a pool of deepest green where cave-divers can follow an open route down 16 feet and along for 65 feet before surfacing in further inclined and jagged passages.

Yet the most surprising find in Claonaite was not in the streamway at all. At the apex of a great heap of shattered rock a wide, muddy opening led upwards, away from the rumbling water into graveyard quiet. An extremely tight slot at the end was passed, and widened, and astonished cavers discovered a vast ancient tunnel, almost blocked by rolling dunes of sand. As they shuffled through, the sand slowly gave way to rubble, the roof soared and a majestic cavern opened out. Slabs of rock clattered underfoot as the cavern halted before a mighty slope of splintered rock, relic of some prehistoric cataclysm. Even here, ways were found leading into the unknown. The best, Infinite Improbability Inlet, lies beyond the vortex of shingle and boulders, providing 100 yards of interesting scrambles through eyeholes and soft mudbanks, gradually getting smaller and smaller and more and more blocked. The end is the most remote underground spot in the Highlands.

In common with other Sutherland caves, Uamh an Claonaite possesses a distinctive character, reflecting the barren moorlands above. Passages wander through dappled, corrugated rock; bits of heather swirl along with the clear water, sidling round cobbles and shingle in a tortoiseshell-coloured streambed, arranged and constantly rearranged by the powerful current. Above all there is a pervading tang of peat, a rich acidic aroma. The water is flavoured with it, the walls are patchy black with it, the floors are soft and yielding with it.

Across the Bottomless Pillar Pool in the main streamway, a series of rocky portholes allows access to a hands-and-knees slither over just such black and slimy peat, followed by a hair-raising

Above An arched passage way in the Viaduct Series.

flat-out squirm in icy water beneath boulders jumbled like building blocks, loose and threatening. Beyond, low boulder-strewn passages run down to a miserable muddy pond and a blank wall, only passable by ducking completely underwater. Strange sights await the caver here. Chambers floored with quivering mud; huge mounds of pure sand, almost filling the otherwise spacious passage; fragile stalactite straws clinging tenuously to wedged, calcite-concreted boulders. These are the rewards of persistence in the little-known byways of Uamh an Claonaite.

Golden eagles soar in the gusty sky above Allt nan Uamh glen; sheep and deer graze on the heathery slopes. But for the caver the real world lies unknown, beneath his feet. The cave's water resurges at the Fuarain Rising. The far end of the known cave doubles back and heads in quite the wrong direction for this rising. Somewhere there must exist a more tremendous stream passage that will delight future generations of cave explorers. This was the first cave in Scotland to pass the mile mark. So much promise does it hold that cavers are already busily looking for the second mile. And perhaps even a third.

SHOW CAVES

1. Poldark Mine
2. George and Charlotte Copper Mine
3. Kents Cavern
4. Kitley Caves
5. Beer Quarry Caves
6. Cheddar Gorge Caves
7. Wookey Hole Caves
8. Clearwell Caves
9. St Clement's Cave
10. Chislehurst Caves
11. Scott's Grotto
12. Hell-Fire Caves
13. Royston Cave
14. Grimes Graves
15. Big Pit Mining Museum
16. Dan-yr-Ogof Cave Complex
17. Dolaucothi Gold Mines
18. Gloddfa Ganol Slate Mine
19. Llechwedd Slate Caverns
20. Dinorwig Power Station
21. Blue John Cavern and Mine
22. Treak Cliff Cavern
23. Speedwell Cavern
24. Peak Cavern
25. Bagshawe Cavern
26. Goodluck Lead Mine
27. Royal Cave
28. Temple Mine
29. Heights of Abraham
30. Poole's Cavern
31. Chatterley Whitfield Mining Museum
32. Ingleborough Cave
33. White Scar Caves
34. Stump Cross Caverns
35. Cruachan Hydro-Electric Power Station
36. Smoo Cave
37. Museum of Scottish Lead Mining

INVERNESS

ABERDEEN

SCOTLAND

• 35

PERTH

GLASGOW

EDINBURGH

• 37

NEWCASTLE

CARLISLE

NORTHERN ENGLAND

33
32

34

YORK

LEEDS

MANCHESTER

LIVERPOOL

23 21
22
24 27
30 28
25 29
31 26
DERBY

20 18
19

SHREWSBURY

14

BIRMINGHAM

CAMBRIDGE

**WALES
AND THE BORDERS**

**SOUTHERN
ENGLAND**

17

13

16 15

8

10

11

BRISTOL

LONDON

CARDIFF

12

6 7

DOVER

9

PORTSMOUTH

SOUTHAMPTON

THE SOUTH WEST

5

2
PLYMOUTH 3
4

1

THE SOUTH-WEST

A mild climate; a lush and gently contoured landscape; superb beaches and coastline: a formula which, hardly surprisingly, attracts great numbers of visitors to the south-west. Underground sites of particular interest are distributed throughout most of the seven counties, excellent excursions for when the pleasures of the beach begin to pall, or rain stops play – or, indeed, in their own right.

In Somerset, the Mendip Hills constitute one of the major British caving regions. On Mendip there are few systems with deep shafts: the emphasis here is on horizontal cave development, often with an active streamway and a fair number of sumps – natural obstacles formed when the cave roof dips below the surface of the water. As a consequence, cavers on Mendip have been responsible for many of the developments over the years in cave-diving techniques. The Mecca for cave-divers for several generations has been Wookey Hole, at the foot of Mendip, and the initial sections of this fine cave are open to the public. Five miles to the north-west a river, now hidden even deeper, has carved down through the limestone over many millennia, leaving the spectacular sheer-sided Cheddar Gorge. At the foot of its great cliffs – which attract the attentions of the country's best rock-climbers outside the busy summer months – is the very popular Gough's Cave. A few yards lower down the gorge is the companion show cave of Cox's – shorter, but very well decorated with a good variety of formations.

Cave development in Devon is on a smaller scale than on Mendip; the longest cave has about 9000 feet of passages, and the great majority of caves have been found in quarrying operations. Many important archaeological excavations have been made in Devon caves, and Kents Cavern at Torquay – a show cave described in full in this section – has yielded many rich deposits. The other show cave in this county, Kitley, has been arranged with a view to providing a sound speleological account, actually on site, of cave formation, decoration, and the like.

Sea caves are frequent in many parts of the south-west's long coastline. Other underground show places reflect the immense importance of mining and quarrying in this region, and the understanding by today's managements of the ever-growing public interest in carefully preserved and well-presented underground workings. The region provides a tangible account of the story of mining covering 2000 years, from the Ancient Iron Mines at Clearwell in the Forest of Dean to the copper workings of the George & Charlotte Mine and Devon and Cornwall's Poldark Mine, where tin provided the incentive.

Poldark Mine

Location: Cornwall
3 miles north of Helston on B3297

Open:
Daily, July and August, 10.00 am to 10.00 pm
September to June, 10.00 am to 6.00 pm

The story of the development of this, one of the most popular tourist attractions in Cornwall, is worth recounting briefly. After Peter Young had retired from a lifetime in the Royal Marines in 1967, he went to an auction in Cornwall to buy a bedroom suite, but instead found himself unable to resist buying a forge in Wendron for £100 freehold! His plan was to open a small craft shop, and adjacent land was later purchased on which to display veteran machines to attract visitors. In the event, the machines became the principal attraction.

Young noticed that the depth of two small pools on the site remained constant, regardless of the weather. The only explanation he could think of was the close proximity of a forgotten old mine, and he began investigating. The underground workings he found were those of a tin mine, the proper name for which is Wheal Roots, dating back to about the beginning of the 18th century.

The tour starts with an audio-visual presentation lasting about eight minutes. Then comes the walk into the drift mine itself. Various viewpoints of particular interest have been identified and the first of these is reached after about 120 feet, where a mineral lode is clearly visible in the roof. Here it is mainly chlorite with some tourmaline and tin, and the orange layer is decayed quartz.

Backtracking a few yards, you turn sharp right into a passage which turns again to take you past the foot of the tilted stope where the miners worked the mineral right to the surface. The passage you follow from this is an adit cut to remove water from the workings. Fifty feet past the stope, continuing straight on at the tunnel crossroads, you come to the foot of the inclined Central Shaft. It was down this shaft (which was cut following the chlorite lode) that Peter Young first entered the mine. It has since been cleared, and gives you the last view of daylight for some time.

Soon the passage turns sharply to the left, and we follow it along a straight line (the miners drove the adit along a fault line in the rock) for over 200 feet. Here the tour descends to a lower level for the parallel route back. But first there is a small detour to another stope in the oldest part of the mine and a view

Below The visitors gaze into high workings in Poldark, the 18th century Mine.

Poldark Mine

up the shammeling shaft, with its platforms and bridges. Men were positioned on ledges every 6 feet up this shaft, shovelling the tin ore up one by one, for Wheal Roots pre-dates steam power.

Back to the adit – the continuation of which is still choked with mud and debris – and a glance up the East Shaft before taking the iron stairway leading to the lower level. Here, at a depth of 125 feet, is a mining tableau, with the tin lode clearly marked by white lines, and a view of a higher level visible through timbers. The passage now rises steadily up an inclined way for 130 feet to a small chamber containing the Poldark Post Box and a collection of old pumps. Passing a short blind passage on the left, you come to the mine's main tin lode, again with the vein marked by white lines. This lode yielded 22lb of tin for every ton of rock removed.

A circular detour to another small chamber now follows, in which stand the stamps used for crushing the ore until the grains of tin could be extracted, a task often done by women mine-workers – the bal maidens. Having looped back to pass close by the main lode chamber, you start the outward journey by crossing a bridge from which you can look down to the long adit followed on the way in. The exit from the mine is along a passage where the ground has deteriorated, and the planked roof is supported by massive timber and metal props.

On the surface, the museum contains a selection of minerals and mining relics from this and other mines. Two are of particular interest. The wheelbarrow in the centre case was found near the Central Shaft, completely submerged but in a state of excellent preservation – thanks, it seems, to minerals in the water. Letters carved on it showed that it came originally from the Wendron Consols Mine, worked forty years *after* Wheal Roots closed down – a puzzle which might be explained by the fact that Wheal Roots was used to drain water from Consols. The other significant, though much smaller, exhibit is a beer bottle seal. Lack of 18th-century records made a proper dating of Wheal Roots difficult, but the discovery of the seal – which clearly bears the words 'John Jane, Wendron, 1735' – provided the necessary clue.

Also on the surface are the many ancient machines assembled by Peter Young. Of particular interest to mining enthusiasts is a Cornish beam-engine which first saw service pumping the Bunny Mine at St Austell in about 1850, and a hoisting-engine designed to be broken down into components small enough to be taken down mine shafts and then re-assembled underground.

Although the section of the mine open to visitors is only a relatively small part of the original 3½ miles of workings, it is steadily being increased.

Wheal Roots connects directly with the adjacent Boderluggan Mine, and at the time of writing work is in progress to open that up too.

Above The entrance to Poldark Mine.

Right Descending into the depths of Poldark Mine.

Kents Cavern

Location: Devon
Just off Ilsham Road, Torquay

Open:
Daily (except Christmas Day),
April to mid-June, 10.00 am to 6.00 pm
Mid-June to mid-September, 10.00 am to 9.00 pm, Saturdays 10.00 am to 6.00 pm
Mid-September to end October, 10.00 am to 6.00 pm
November to March, 10.00 am to 5.00 pm

Although in a seaside town, and not that far from the sea front, Kents Cavern is not a sea cave. Its formation followed the inland pattern, of solution of passages by down-flowing fresh water, and not that of wave action. Apart from its reputation as a well-decorated tourist cave, Kents has interested scientists for more than 150 years because of its rich yield of human and animal remains. Amongst the many boulders littering the floor of the cave, archaeologists have found, embedded in a black mould, remains dating from the present day back to the Neolithic. Flint implements discovered buried *underneath* a calcite floor have been dated at least 100,000 years old. The skull of a young woman, found by the cave's owner lodged in a rock crevice outside the entrance, was dated to some 20,000 years.

Kents Cavern has over ½ mile of passages and chambers, most of which are seen on the tour. You start by examining the first opening in the cave wall, the Hyaena's Den, where bones of that prehistoric animal (much larger than today's hyaena) were found, then follow a gentle slope to the entrance of Charcoal Cave on the left. In the small passages opening up from this cave were found the remains of fires where Stone Age families had their home and

working place. (The exit from Charcoal Cave will be seen just before you leave by the South Entrance.)

On the right of the passage, opposite the entrance to Charcoal Cave, is the very deep excavation where archaeologists discovered one massive tooth – weighing 7lbs – of a mammoth now displayed in the entrance hall. As it was impossible for a mammoth to enter the cave, researchers assume that it was killed close to the cave's entrance and that the tooth was washed in by the river that flowed then.

Now follows the aptly-named Long Arcade, some 200 feet long, the roof soaring to a cathedral-like arch in places. After a glimpse of the entrances to Underhay's Gallery and Little Oven comes a flight of six steps leading up to the higher level and the large stalagmitic formation known as Hedges Boss.

Two detours lead off from Hedges Boss. The first passes along Clinnick's Gallery to the Organ Chamber, finely decorated and displaying a row of stalactites draped in the manner of a church organ's pipes. Beyond lies even greater splendour, in Rocky Chamber, or Stalagmite Grotto. Here, at the deepest point in the cave, lies a wealth of stalagmites and stalactites; the three-tier Wedding Cake and the Chinese Pagoda are particularly outstanding. Of especial interest is the stalagmite growing behind a glass security screen. Despite its considerable height of 54 inches, it is remarkably slender for a

stalagmite, which indicates a very slow growth rate: 50,000 years is suggested.

The second detour from Hedges Boss is into the spacious Cave of Inscriptions with its delicately-shaded Waterfall and large stalagmite boss. Carved into this is an inscription. While this form of graffiti is hardly encouraged by show-cave owners nowadays, the antiquity of this one makes it of particular interest: 'Robert Hedges of Ireland Feb 20 1688'. Surprisingly, Hedges was traced without much difficulty – not to Ireland the country, though, but to a small farm bearing that name just a few miles from Torquay.

From here begins the outward journey via the Labyrinth to the Bear's Den, situated at the start of the second of the cave's two parallel passage developments. Here you can see the original floor level, high above the passage leading out, and the floor of bones – a limestone base rich in embedded bones of long-extinct animals. Two specimens are seen: the skull, with jaw-bone, of a great cave bear, and just the tip of a 7-inch tooth from a sabre-toothed tiger (the rest was successfully recovered and is now in the Museum of Natural History in London).

The tour ends, via the South Entrance, after the final spectacle of the Great Chamber, where a floodlight is turned on to illuminate a final display of coloured stalactites and a natural rock formation, the silhouette of which gives rise to its name: the Camel.

Above Rocky Chamber or Stalagmite Grotto.

Right Stalactite cascade in Kents Cavern.

George & Charlotte Copper Mine

Location: Devon
Morwellham Quay, Morwellham,
Tavistock; bear left off A390 2 miles
west of Tavistock

Open:
Daily (except Christmas Day),
1 March to 30 September, 10.00 am
to 6.00 pm (last admission 4.30 pm)
1 October to 28 February, 10.00 am
to dusk (last admission 2.30 pm)

A visit to this mine is just one of the interesting things to do here, the rest of the itinerary being most of the village of Morwellham Quay itself. The village is tucked deep in the Tamar valley, in a designated Area of Outstanding Natural Beauty, hard by the side of the river Tamar. Ships have moored here for over 700 years, but it was during the nineteenth century that the port was at its busiest, when it was one of the most important copper-ore exporting centres in Europe. At that time the Tamar carried more shipping than the Mersey, country lanes leading down to the almost continuous quay along its banks.

Morwellham lies at the centre of the southern edge of the mineral belt between Tavistock and Callington, a belt rich in copper and arsenic. Copper ore has been found within 5 feet of the surface here, and there are outcrops of it on the river bank above and below Morwellham. The quay's busiest time came after the opening of the Tavistock Canal in 1817. Running from Tavistock to Morwellham, the 4½-mile-long canal involved the excavation, lasting fourteen years, of a 1½-mile-long tunnel through the high ground circling Morwellham. The rich copper ore from the mines of Wheal Friendship and Wheal Crowndale near Tavistock could

at last be brought easily to the Tamar instead of using expensive teams of pack-horses. As the canal finished 237 feet above the Morwellham quay, ore was brought down this last steep descent by means of an inclined railway, powered by a waterwheel.

Morwellham Quay's heyday begin in 1844, when one of the country's richest deposits of copper was discovered at Blanchdown, 5 miles to the north. Up to 4000 tons of ore would be waiting on its quays at any one time, ready for transport by the fleet of 300-ton schooners. So intimately were mines and port bound, however, that when their fortunes flagged so did its. The arrival of the Great Western Railway at Tavistock in 1859 took much of the canal's traffic, and it ceased business in 1880. As the copper ore became exhausted, the production of arsenic became the mainstay. The white powder, remembered today chiefly for its more sinister applications by notorious poisoners, was widely used in the manufacture of paints, dyes, insecticides and glass. When that business too ceased, Morwellham Quay slipped into obscurity and disrepair for over 60 years.

The busy open-air museum that can be seen today is the result of the leasing in 1970 of parts of the village by trustees of Dartington Hall and Dartington Amenity Research Trust. The admission ticket gives access to a wide range of attractions, including the quays and raised railways; working waterwheels; a blacksmith's, cooper's and ore assayer's workshops; the canal; two audio-visual shows; the hydroelectric power station (powered by the canal water); several museums and various riverside and woodland trails.

Of particular interest is the train ride into the George & Charlotte copper mine. Some 500 feet of the mine's deep adit have been cleared and opened up, and this, coupled with a *son et lumière* presentation using life-size figures, gives the visitor an idea of the working conditions inside a typical Tamar valley copper mine.

Last worked in the late 1860s, the mine had a twin in the William & Mary mine. Each worked the copper-ore deposits, which extend for nearly 1 mile, on either side of the high ground separating the Tamar from the river Tavy, and they may indeed have been connected at some point. When worked, the greatest deposit of copper ore found here left a huge excavation

known as the Devil's Kitchen; an expensive crosscut tunnel made from Ley's Shaft to try and intercept any possible downward extension of this ore pocket, but apparently without success.

In common with other mines worked in fairly steep hillsides, George & Charlotte had the advantage that, at least as far down as the river level, water entering the workings could be drained off through relatively short adits. This gave a flood-free zone 500 feet high, and only the lower workings – down to the 54 Fathom Level – required pumping to keep them clear.

The mine has proved such a popular attraction at Morwellham that, at the time of writing, an extension is being driven into adjacent old workings, to make the route a circular one.

Above 600 feet below ground, the train arrives at
the *son et lumière* tableau.

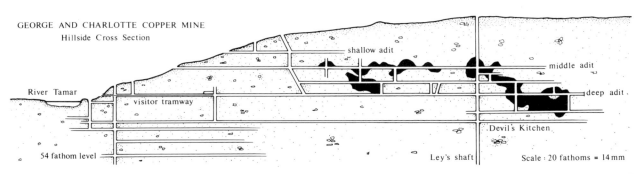

GEORGE AND CHARLOTTE COPPER MINE
Hillside Cross Section

shallow adit

middle adit

deep adit

River Tamar

visitor tramway

Devil's Kitchen

54 fathom level

Ley's shaft

Scale : 20 fathoms = 14 mm

Kitley Caves

Location: Devon
Signposted from A379 ½ mile south-west of Yealmpton

Open:
Daily, Good Friday and Easter week, and from Saturday before Spring Bank Holiday to end September, 10.00 am to 5.30 pm

Though among the smaller show caves, Kitley has been sympathetically developed by its owners, and has the added bonus of being situated in some fine Devonian riverside country. The ¼-mile walk from the car park to the caves is alongside the river Yealm, flowing from the granite moors of southern Dartmoor. Three disused lime kilns are passed and, close to the dismantled railway bridge, two underground streams resurge into the river from holes in the bank. The shallow weir, with its adjacent hut and calibrated post, is where the Cornwall Water Authority measures the varying flow of the river – an average of 36 million gallons a day.

Both salmon and trout can be seen in the right seasons, and more permanent residents include badgers, water voles and minks.

The path leads to the old quarry workings which ate into the cave system here at Kitley. Half a dozen small caves can be seen on the 100-foot-high quarry face, as well as exposed fragments of cave walls remaining from when the blasting tore apart the ancient passages and chambers. On the hillside above the quarry is a long line of dolines – surface depressions denoting a line of cave development *somewhere* underneath. Efforts are being made to dig into this.

Well worth seeing, either before or after your trip underground, is the museum and interpretation centre, sited within the great arc of the quarry. As well as some of the archaeological finds, there are displays giving a clear picture of cave development and of the work undertaken by the Society of Devon Karst Research in excavating for new caves.

The show cave itself is basically a series of small passages leading off from a central chamber. Rather than turning the cave into an illuminated 'wonderland', it has been laid out as a serious speleological exhibit, with notices explaining the various features encountered on the tour.

The entrance is via a passage revealed by the quarrying and the cave has been open to the public, on an irregular basis until recent years, since 1793. To the right, at the first junction, is the Well – the waters of which connect with the river – and Fred's Folly, a determined but eventually unsuccessful cave dig. A turning left takes you along Flood Passage and, after a short distance, into the main chamber.

This chamber houses a number of different cave features. One of the principal ones is the Central Massif, a stolid pouring of flowstone. The path by here was dug through the old water-level deposit. After the formation of the chamber below water level (the clue to which is given by the flat roof), the water table fell, the turbulent flow leaving the erosional features clearly visible on the roof. Mud, pebbles and even some animal remains settled on the floor; then the water receded, leaving the slow calcite drips to seal them over.

The exit passage takes you past a low bank of gour pools – natural dams of slowly deposited calcite holding back curving miniature lakes.

Exploration and excavation continue at Kitley, so future extensions may well be opened. In the meantime, if prior notice is given, trips to some of the less accessible caves can often by arranged.

Left The use of different coloured lighting is a special feature of the tourist trail in Kitley Caves.

Beer Quarry Caves

Location: Devon
Turn south from the A3052 on to the B3174 just west of Seaton, then right into Quarry Lane; the caves lie on the left

Open:
Daily, Easter to 30 September, 10.00 am to dusk
For off-season visits, telephone Seaton 20986

Devon's Beer Stone must, it seems, be the mason's dream material: soft to cut and carve when first quarried, it actually hardens on exposure to air. Its history takes us back some 120 million years to the time when exactly the right proportions of pulverized shell, refined clay and fine sand settled on the sea's bed to weld, over aeons, into this fine-grained and creamy-white mineral.

Not far away, huge excavators claw away the overlying layers of soil and chalk to reveal the still yielding great seam of stone, but in Beer Quarry Caves (opened to the public in 1984) the great chambers, halls and corridors are silent except, perhaps, for the dim, imagined echoes of those countless men who worked the stone here for nearly 2000 years.

The underground workings cover approximately ¼ square mile, and the conducted tour lasts 1 hour. When we explore the past in an archaeological dig, that which is oldest is last to be met. It is always a surprise, then, when entering an underground working, to remember that the very portals are the most ancient part, not the innermost recesses. This adjustment has to be made here, too, for you enter the quarry through a Roman arch where the past is clearly recorded through the many thousands of pick-marks still visible after almost twenty centuries.

Romans were the first to tunnel into the hillside inland of the great sea cliff where they spotted the 12-foot-thick beds of white freestone, realizing its worth in the building of their villas. Norman followed Saxon, extending the original Roman workings, and in the early Middle Ages the versatile qualities of Beer stone were exploited to the full for both delicate carving and rugged engineering, from intricate tracery to massive buttresses.

While we might deplore the activities of those who leave their various graffiti on the walls of a bus shelter, here at Beer the passage of history changes our viewpoint completely, and we welcome the countless signatures of long-dead masons, some dating back to the 17th century. Thanks to these, in some cases several generations of families can be traced.

The smooth working of many of the walls and roof-supporting pillars may seem puzzling at first, but bear in mind that the stone was not blasted out but cut by axe or giant handsaw. The only machinery used was a primitive crane for lifting the stone blocks, each weighing 8 tons on average, on to carts pulled by teams of four or six shire horses. Each block of stone was marked in a particular way at the quarry so that the mason could place it as it had lain in its original bed. This prevented rainwater from seeping into the pores, or grain, of the stone and consequent frost damage.

The use of Beer stone in buildings of special merit has continued almost uninterrupted through the centuries. First used by the Romans in Honeyditches Villa at Seaton, it was employed extensively in building Exeter cathedral from the early 12th century, and can also be found in St Paul's Cathedral and Westminster Abbey. Sir George Gilbert Scott favoured it in the nineteenth century for his restoration work on Wren churches, and it can even be found in the USA in the Cathedral of All Saints in St Louis, Missouri. There is one especially appealing demonstration of the almost timeless nature of Beer Quarry Caves: stone being used in the current restoration of Exeter cathedral was cut from the mother bed only yards away from where the *original* stone was cut.

Right In one of the great carved galleries of Beer Quarry Caves.

Cheddar Gorge Caves

Cheddar Gorge Caves
Location: Somerset
On the B3135, immediately above Cheddar village

Gough's Cave
Open:
Daily (except Christmas)
Summer, 10.00 am to 6.00 pm (last tour starts
5.30 pm)
Winter, 11.00 am to 5.00 pm (last tour starts 4.30
pm)

Gough's Cave, Adventure Caving
Available:
Daily (except Christmas), normally at 9.30 am,
11.30 am, 2.00 pm and 4.00 pm. Advance booking
strongly advised; telephone Cheddar 742343

Cox's Cave
Open:
Daily (except Christmas)
Summer, 10.00 am to 6.00 pm (last tour starts
5.30 pm)
Winter, 11.00 am to 5.00 pm (last tour starts 4.30
pm)

Fantasy Grotto
Open:
Daily, Easter to mid-October, 10.00 am to 6.00
pm (last tour starts 5.30 pm)

Cheddar Gorge, a sheer-sided gash for much of its 2-mile length, is unequalled in Britain. Its soaring limestone cliffs, up to 400 feet high and presenting the rock climber with some of the steepest and most demanding climbing in the country.

Its origin is disputed, but most geomorphologists now seem agreed that the gorge was probably hewn by a river which has since quit the surface and now finds its way underground through unknown caves.

The sides of the gorge are peppered with various caves – including, perhaps, the mysterious Lost Cavern of Cheddar, mentioned by the historian Henry of Huntingdon in the 12th century. He wrote of 'Cheder Hole, where is a cavity under the earth, which, though many have often entered and there traversed great spaces of land, and rivers, they could never yet come to the end', but the 'Hole' has never been positively identified or located. There is no evidence that the great caves of Cheddar were entered between pre-historic times and the last century, so the mystery remains.

Most of the gorge's caves are small and the domain of the experienced caver, but at the bottom end of the gorge, where it runs into the village of Cheddar itself and dissolves in the southern flanks of Mendip, are the famous show caves of Gough's and Cox's, plus the very much smaller, and quite artificial, Fantasy Grotto.

Gough's Cave

Gough's, the topmost of the gorge's three show caves and the largest, is probably the one to choose if you only have time for one visit. It offers a good cross-section of natural passages, chambers and formations, and the lighting is sympathetic for the natural coloration.

The entrance would be hard to miss, situated at the foot of the complex of concrete buildings housing a museum, exhibition, information centre, gift shop, restaurant and cafeteria. Do make a point of visiting the museum, the captivating centrepiece of which is the skeleton of Cheddar Man. Various diorama give an impression of Man on Mendip through the ages, and a series of illustrations shows the development of Cheddar Gorge and its caves. Various of these caves have been occupied through history and pre-history, and there are artefacts dating from the Early Upper Paleolithic (30,000 to 20,000 years ago) right up to the Iron Age (roughly 550BC to AD43).

Inspired by the success of Cox's Cave in the gorge, Richard Cox Gough and

Above The flat surface of a dark pool provides this enchanting mirror effect in Aladdin's Cave, Gough's.

Left Cheddar Gorge – a great, steep-sided gash through the limestone of the Mendip Hills.

his sons began excavating the inner reaches of a short cave at the base of the cliff in 1890. The fine passages and chambers of the cave were only slowly revealed over a period of eight years to the Goughs. Their eventual reward was the discovery and opening-up of one of the country's best-known show caves.

The tour begins in the ruggedly arched and wide passage which descends past the Skeleton Pit. It was in this pit that, only a few years after the opening of the cave, the 10,000-year-old skeleton of Cheddar Man was found. The clearly visible damage to the front of the skull suggests that he was killed by a blow to the face.

The first main formation you come upon is the Fonts, on the right-hand side, a succession of beautifully sculpted ascending gour pools: natural dams formed by slowly deposited calcite from the gentle water flow. This is a good example of how cave decoration demands quite different levels of flow from the formation of a cave, very often a turbulent affair. Softly, softly is the key here.

The water-filled crevice near here has been dived to a narrowing, too small to pass, at a depth of over 60 feet. It is from here that the cave water gushes in time of flood, even reaching the roof in extreme conditions – but, you will take comfort to note, this takes some time to happen.

Continuing up the Grand Passage you reach the gentle incline of Heartbreak Hill. Gentle it might be when walked up or down, unencumbered, but it takes its name from the efforts of those who had to steady many hundreds of wheelbarrow loads of cave-fill down it from excavations beyond. The author has reluctantly indulged in this pastime himself, carting down spoil from a caver's dig leading off the hill, with only the prospect of virgin passage making up for the mad pell-mell careering. Part of the roof here is patterned with the shell-shaped scoops called scallops, formed by localized underwater swirls in the river that formed the cave.

At the top of Heartbreak Hill are what remains of a group of stalactites with the telling name of the Ring

Cheddar Gorge Caves

Left Frozen River, Gough's Cave.

o' Bells. With only two now unbroken, tunes are no longer tapped out by the guides, as one suspects they were in the old days. Take particular note of the small pool at this point: this marks the original floor level before all those barrowloads of deposited mud were trundled down Heartbreak Hill to give easy access.

Beyond the now silent stalactites, steps lead up to a passage which was, until the mid-1970s, a dead-end. To create a circular route through this, the narrower part of the show cave, a tunnel was blasted for 120 feet to the splendid chamber of St Paul's. Here, down one complete wall, tumbles and spills a huge and finely coloured flow of calcite, softly rounded protrusions capping finely fluted ribs. Beneath the great cascade is the tiny grotto of Aladdin's Cave, its stalagmites and flowstone carefully illuminated and reflected in the artificial pool.

A short walk leads to the show cave's second spectacular chamber, King Solomon's Temple, again with one wall a vast calcite cascade. Above is spotlighted a richly stalagmited chamber, and lower down the deliciously creamy coloured and textured Frozen River and Niagara Falls.

Now begins the return, and the guide will point out, on the right, the trick of light and shadow known as the Black Cat. The non-show cave continues past this point for some little way to a drop into a chamber where David Lafferty stayed alone for 127 days in 1966 – a world record which lasted only a short while.

A long flight of steps takes you down to rejoin the Grand Passage. At the bottom, look up for a rather surprising sight so far underground – a small 'garden' of Hart's Tongue ferns, succoured only by the light from the bulbs, and established in this alien setting quite by chance.

As you make your way out along the Grand Passage, there are two final natural displays. The wide, low section carpeted with a large cluster of subtly-coloured stalagmites is the Pixie Forest. And, on the opposite side, lies Swiss Village – a group of stalactites and curtains mirror-imaged in the water of the placid pool beneath.

Gough's Cave – Adventure Caving

Opportunities for non-cavers to participate in a proper caving trip usually involve contacting a local caving club or attending a course at an outdoor activities centre. Here, though, is the chance to get at least a taste of 'wild' caving, far removed from the floodlights and concrete paths of the show caves.

At the entrance to Gough's you will be provided with a helmet, boiler-suit and miner's electric lamp. You should turn up, fifteen minutes early, in your oldest clothes and a pair of stout boots or wellingtons. The only other requirements are that participants should be over twelve, reasonably fit, and not subject to epilepsy or dizzy spells.

A detailed synopsis of the trip – which takes between 1 and 1½ hours – would be inappropriate, for the essence of caving is discovery. But the trip, which takes you through some of the non-tourist parts of Gough's, is designed to show a fair cross-section of the types of obstacles encountered in the sport and the techniques used to surmount them. Parties (limited to a maximum of ten people) are led by an experienced caver. As well as guiding you through the important topics of safety and conservation, he will give a fascinating insight into the formation of caves and the sport you are experiencing for the first time.

Cox's Cave

The entrance to Cox's is a short distance down the gorge from Gough's and on the same side. Despite being much smaller than Gough's – its total length is only just over 300 feet compared with Gough's 3750 feet, though not all of that is open to visitors – its formations are first class and, in many cases, closer to the visitor. If Gough's is the boudoir, Cox's is the jewellery casket.

The cave was discovered in 1837 by George Cox, that much is known for certain, but exactly how is less sure. Some have it that he was quarrying limestone opposite his water mill (now a hotel), others that he was digging into the rock face to make more room for his carriages, while yet another, more prosaic, explanation is that the road was being widened. Whatever the reason for the excavation, it was a stroke of fortune for Cox when that first small hole was struck into the

Cheddar Gorge Caves

blackness of the cave void. Astutely he realized the commercial value of his find, and wasted no time in opening the cave to visitors.

About half-way down the cave an area of scalloping provides a useful indicator of the direction of water flow when the passage was formed. Close examination of the scallops will show that one end is deep while the other tapers to the rim of the next depression. Knowing that the deep end is formed downstream, we can tell that the river followed the direction taken by the visitor into the cave.

The general layout of Cox's is that of seven small, well-decorated chambers joined by low archways. Of the main formations, some of the most memorable are those in the Transformation Scene, where yet again (a favourite Cheddar device, this) a pool mirror-images stalactite columns; the Speaker's Mace – a tall stalagmite which has decided to burgeon out at the top rather than the base; and the Mermaid and Mummy, a chamber in which the etched and scooped limestone walls have been partly covered by brightly coloured calcite. The whole

effect is more that of some fantastic whale's innards than of a cavity in the earth's depths.

In the heavily decorated chamber which follows the Entrance Chamber, the discoverer was faced with a problem. So close together were the formations that Cox could foresee difficulties in getting visitors through this section, and so he dug underneath the grotto. As a result, the decorations still stand in their original glory.

At the deepest point in the cave – under the beautiful Marble Curtain – is a pool. This is said to follow the varying height of the river outside, but always 10 feet lower, so its link with the river must lie downstream. Steps take you back to the surface, via a second entrance, passing the stalactites, stalagmites and columns of the Ladye Chapel.

Fantasy Grotto

Mendip cavers have a saying: 'Caves be where you find 'em.' And if that means digging for month after month, or year after year, that is just what you do. It seems that this philosophy goes back a long way – to the end of the last century, in fact. Then, Roland Pavey acquired the land next to Cox's cave, and became determined to find *his* own show cave. The resulting 'cave' he dug and blasted over the years, taking advantage of several natural open bays in the cliff face which he walled off from the persistently intruding sunlight outside. From its opening to the public in 1890, this excavation fought a rather one-sided battle with the natural underground wonders just a few yards up Cheddar Gorge. Even the old ploy of changing the name was tried – Pavey's Cave, *Miss* Pavey's Cave, Aquarium Cave, Cave Aquarium, Waterfall Cave . . . The present management has taken a more realistic approach, and has populated the grotto with King Neptune, sea monsters, pearl miners, mermaids and the like. Their animations provide a quite different attraction for those children too young to appreciate the 'boring old stumps and spikes' of Cox's and Gough's.

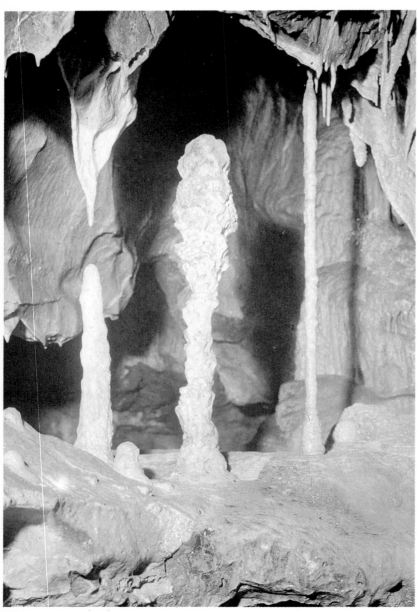

Left The Speaker's Mace, Cox's Cave.

Right The Mermaid and the Mummy, Cox's Cave.

Wookey Hole Caves

Location: Somerset
Wookey Hole village (*not* Wookey),
2 miles north-west of Wells

Open:
Daily (except Christmas Day),
March to October, 9.30 am to 5.30 pm
November to February, 10.30 am to 4.30 pm

Wookey Hole is a place enormously evocative of the past. While not endowed with the richness of formations of some other show caves, its chambers are impressive, and its importance in local history, pre-history and legend is considerable. And for British cave divers, it is their Mecca . . .

A pleasant walk from the car park, past the old paper-mill complex, brings you to the entrance near the foot of a 120-foot high cliff. From its base flows the river Axe, its volume varying by the season, from 1 million gallons a day in summer drought to 40 millions gallons in the wetter winters.

As you go into the entrance, you are following in the footsteps of those for whom Wookey Hole was a place of security, a home, 2000 years ago. Many painstaking archaeological digs over the years have shown that the cave was regularly used as a habitation from 250BC until about AD450. Yet there is something puzzling and disturbing about these dates. Why, when other, shallow, caves nearby are known to have sheltered people for possibly 50,000 years, did the occupation of Wookey Hole come so late? And, equally puzzling, why did it last for only a few hundred years? We know today that at certain water levels strange and startling noises come from the flooded

sections of the cave. Did these frighten off potential dwellers, or merely reinforce some existing darker taboo?

As you walk the 300-foot passage which takes you from daylight to the 1st Chamber, you pass the spot where the remains of the cave's last inhabitant were found. In 1912, Herbert Balch excavated here; he was one of Mendip's leading amateur archaeologists and early cave explorers, many of whose finds can be seen in Wells Museum. Balch unearthed the skeleton of an old lady, who died, apparently of natural causes, in the Anglo-Saxon period before the Norman conquest. Next to her were the remains of her two goats, tethered to a stake, plus a milking bowl, a comb and a crude knife, and – more enigmatically – a cut and polished ball of stalagmite crystal.

Here was, almost certainly, the remains of the woman who had come to be known over the centuries as the infamous Witch of Wookey. More open to interpretation are other finds: the skull of a teenage girl, savagely smashed by a blow on the top, and a mass of human bones broken as though for the cooking-pot.

Let us pass, probably hurriedly, along the passage from the entrance – where villagers used to leave gifts of precious pins for the witch – to the 1st Chamber, via the stairs of Hell's Ladder. In this place, 150 feet long by 50 to 60 feet high and wide, often given the alternative name of the Witch's Kitchen, we have our first sight of the underground river Axe. We see what appears to be a blue, calm lake; the great thrust of the river is deep, entering from the submerged arch clearly visible beneath the flat surface.

Gazing across the river is the large Witch stalagmite. Here, legend tells us, she was turned to stone (as was her dog), after a long battle with a monk from Glastonbury. Perhaps for once, though, the hard facts of the archaeologists' finds are more disturbing than the legendary 'frozen figure'.

The wooden boat on the lake was brought into the cave in 1946 by the early cave divers to hold their helpers. The group of stalagmites behind the Witch includes the White Giant – 9 feet

Right The dark silhouette of the petrified witch gazes across the waters of the River Axe in the 1st Chamber.

Wookey Hole Caves

high and 40 feet around the base – and, in a pool where Celtic pottery was discovered, an island stalagmite named St Michael's Mount. Close examination of the straws above the Witch shows that they bear testimony to the sudden change from candles and oil lamps to electric lighting in 1927. One or two inches of pure white calcite tube have already grown below the clearly visible soot line, now incorporated into the crystalline structure.

Now we pass through the short, but 70-foot high, 2nd Chamber – with a spur of the river Axe making an appearance on the right and a number of flow-stone formations – to duck under a double arch into the impressive 3rd Chamber: the Witch's Parlour.

Although only 14 feet high, the domed 3rd Chamber is a memorable place, almost circular and 100 feet in diameter. Here again the river Axe occupies much of the floor area, and yet again the impression is of a placid and

lucid pool. But when the guide extinguishes the main lights, you will see clearly the great arch through which the river pulses, illuminated from the 4th Chamber in the flooded beyond.

The 4th Chamber has been cut off since 1857, when the water level was raised by the building of a dam to feed the leat to the paper mill. In 1935, Graham Balcombe marked the beginning of the cave-divers' era in Wookey by passing through the archway to the 4th. Excavations showed that the Romano-British inhabitants had used the 4th Chamber for burials, bearing the bodies aloft as they waded in through the river.

For many years, the 3rd Chamber was the departure-point for cave divers intent on exploring the river ever further to its sources in the caves on the Mendip plateau above. In 1936 Balcombe, accompanied by Penelope Powell, discovered and explored the 7th Chamber, giving a running commen-

Above Wookey Hole – for many years, this, the 3rd Chamber marked the end of the tourist route and the start of the cave divers' domain.

Right An excavated tunnel now reveals the glories of the 9th Chamber.

tary for the BBC as he did so. Both wore the standard naval diving-suits of the time, heavy and cumbersome with their copper helmets and rope and airline to be dragged behind – a far cry from today's wet-suits, face masks and self-contained breathing apparatus.

Until 1975, this was as far as the public could go, but an ambitious tunnelling project was then undertaken, opening up the 7th, 8th and 9th Chambers to visitors. The first two are very similar in character – lofty and long, but narrow, crossed by bridges 20 feet above the deep water. Here there are ochre-coloured flowstone formations, with patches of soft milky white.

Wookey Hole Caves

The 9th Chamber is on a much grander scale, rising nearly 100 feet to where passages lead up further to the very surface, though this way is now sealed. The orange cord in the water has been laid by cave-divers, and provides their only means of safe return when the river-bed silt has been disturbed by their progress, reducing visibility to virtually zero.

With the easy access to this part, all diving operations now use the 9th as their base. Since 1935, Wookey Hole has been the focal point for British cave divers, and has become, in effect, their 'Everest'. In 1976 ¼ mile of submerged passages were passed, yielding Wookey 24 – nearly ½ mile of dry cave. Then followed another demanding sump, 80 feet deep and 300 feet long, before the chamber of Wookey 25.

The dive from this point takes cave-diving to its extreme limits. At the time of writing, the greatest depth reached in this sump is 200 feet, achieved by Martyn Farr – the leading British exponent of this most demanding form of cave exploration – in 1982, during a diving trip lasting 24 hours in all.

The visit into the cave ends with the remaining few hundred feet of tunnel that take you back to the surface, and the sight again of the river Axe coursing out from the rock, its ultimate underground secrets still intact. The tour concludes, on the surface, with the displays in the old paper mill of old fairground attractions, a demonstration of paper-making by hand, hundreds of stored wax heads from Madame Tussaud's in London, and a cave and cave-diving museum.

Above The waters from the River Axe escape to the surface after their long limestone confinement.

Right The Witch's Kitchen in the 1st Chamber. Her dark silhouette is visible in the lower right-hand corner of the picture.

Clearwell Caves

Location: Gloucestershire
1½ miles south of Coleford, just north of junction between B4231 and B4432

Open:
Daily except Saturdays and Mondays, Easter to 30 September, 10.00 am to 5.00 pm
Winter season by appointment; telephone Dean 23700

Bounded by the Wye, the Severn and the Herefordshire border, the Forest of Dean is a blessedly overlooked part of Britain. Yet it contains one of the greatest primeval forests in the whole country — millions upon millions of oaks, hollies, birches, ash and conifers. Beneath the canopy of leaves spreads a network of pathways totalling, some say, 2000 miles.

But the wealth of the region has lain beneath the roots of the great forest, in its 22,000-acre coalfield and in its iron-ore deposits. Since pre-Roman times the minerals of the Forest have been worked, not by huge mining conglomerates but by the Free Miners of the Forest, often in small mines employing just a handful of men. That the iron mines were worked well before the Roman invasion is borne out by the many ancient terms still used even today which have Celtic rather than Latin roots. Examples are, 'scowles' for the open-cast workings; 'gales' for the miners' allotments or holdings; 'crease', the name given to the limestone here. The many strange laws and

Right A mining track on its 14″ gauge track together with other implements in Clearwell Caves.

Clearwell Caves

customs which still abide, although stretching back tens of centuries, are unique to the Forest.

At Clearwell Caves, the visitor can gain a first-hand impression of the labours involved in wresting the ore from the Forest deeps. And, of course, charcoal from the countless trees made the Forest an ideal place for the smelting of ore as well. The iron for most of the English weapons used in the later Crusades came from here, and over the last 54 years of the 19th century a total of 62,000 tons of ore was extracted from Clearwell alone. Even as late as the First World War, it yielded another 3000 tons. More recently mining has been on a much smaller scale, only red ochre being extracted for paint.

The great age of the place is brought home even before you enter the caves, for the two outcrops of rock (King and Queen Stones) before the entrance mark scowle workings – the most ancient type, open to the surface.

In through the entrance, where t'owd man (as the old miners are affectionately known) first noticed and began to follow the iron ore, one reaches a chamber. Take the left-hand passage, but then stop and look carefully along the railway to the left. We know, of course, that the silhouette of the old man's head is really that of a rock formation, but there are those who know it equally well to be the petrified remains of an ancient wise man who was foolish enough to upset the spirits of Nature.

The clarity of the pools of water hereabouts is quite marked, and it was to these pools that the inhabitants of nearby Clearwell Meend came for their drinking water until as recently as a few dozen years ago.

Bat Chamber, opening off to the left, offers secure hibernation quarters for a variety of bats, the constant temperature of 50 degrees Fahrenheit giving life-saving warmth during the winter season. The white formation in the White Lady Chamber at the end of this passage is – though we should have guessed by now – a petrified princess, the continuing search for whom caused the petrification of the old man glimpsed earlier.

Rejoined, the passage leads you now into the large Old Churn chamber, 140 feet long and 80 feet wide. On the right-hand side you can still see pick marks left by the miners as they won almost every last scrap of iron ore – and all by

the light of a candle supported on a stick clenched between the teeth.

A low winding tunnel now leads through to Chain Ladder Churn where the miners climbed down to the mine's lower regions, extending another five times as far. The chamber is just as t'owd man left it, down to the traces of fire setting, an alternative method of mining to the pick. A fire was lit to heat the rock, which was then drenched with water, causing it to crack.

There is a delicious, if spooky, story associated with this chamber. A tele-

vision crew, filming in the caves, discovered that their cables would not reach. An old man appeared, suggested what proved to be an ideal alternative route, then vanished. None in the caves could even guess at his identity!

The name of the next chamber, Barbecue Churn, can hardly be said to date back into the mists of time. The great size of the chamber makes it ideal for various public functions – including, of course, barbecues. Much ore was mined here, then it was used for dumping spoil from other sections so

that the original floor is now buried beneath 15 feet of rock. Most of the flat roofs in the cave – like the one visible here – are of Whitehead limestone, which the miners called lid stone; the ore was found in the band of Crease limestone, while the floor marks the top of the Lower Dolomite limestone.

Pillar Churn – reached by a 'new' passage cut in 1885 – takes its name from a column of stone, and a sharp left turn brings us into Frozen Waterfall Chamber, with a calcite flow spilling down. The last chamber of the tour,

Pottery Pocket, was once open to the surface, and its claim to fame now is as a rubbish tip! Many intriguing artifacts – some dating back to the fourteenth century – have been unearthed here.

From here a winding passage takes you back to the surface, via the Engine House, which contains a variety of vintage engines and mining equipment. But cast a thought back to t'owd man who, without the benefit of such mechanical help, had to hump to the daylight every fragment of precious ore in a wooden box strapped to his back.

Above The Entrance Chamber, Clearwell Caves, where visitors walk through mined galleries, which provided much of the iron for the English weapons in the later Crusades.

SOUTHERN ENGLAND

Although the term 'cave' is used to describe no fewer than five of the entries in this section, natural caves are few and far between in the region and are generally restricted to fissures in the chalk, particularly along the south coast. Yet it is an area in which man has delved underground for a variety of reasons.

In at least three of the sites quarrying was the main motive: the maze of Chislehurst Caves in Kent, the notorious Hell Fire Caves at West Wycombe in Buckinghamshire, and the 4000-year-old flint mines of Grimes Graves in Norfolk. St Clement's Caves in Sussex may also have come about through no more than a straightforward quarrying operation to extract sand for the glass-making industry. Mystery there is, though, over the origins of the puzzling Royston Cave, which may have had links with the Knights Templar.

Although cavers living in the south have to travel to other regions to find cave systems of any size and challenge, they often sharpen their teeth, so to speak, exploring the multitude of man-made sandstone and chalk excavations in the southern counties of the area. Indeed, London cavers (particularly members of the Chelsea Speleological Society) are frequently called in by local authorities to explore and survey ancient workings which suddenly come to light, usually in embarrassingly close proximity to a road or a housing estate.

St Clement's Caves

Location: Sussex
Croft Road, Hastings

Open:
Daily, summer, 10.15 am to 12.15 pm, 1.30 pm to 5.15 pm
Winter (according to demand), weekends, tours at 10.15 am, 11.15 am, 12.00, 2.15 pm, 3.15 pm, 4.15 pm

These caves, which cover about 3 acres, are far removed from the decorated natural limestone caves of other regions, but – with their multi-coloured lights and abounding legends – they make a pleasant and out-of-the-ordinary retreat, especially on a hot summer's day. Formed in soft white Ashdown Sandstone, the vaulted passages and chambers have obviously been largely made by human hand. Compared with the relatively geometric excavations of Bo-Peep Cave in nearby St Leonards' sandstone, St Clement's Caves do follow a more complex and random pattern, suggesting that natural cavities in the stone were enlarged artificially in the past. Some maintain that sand was mined here for the glass-making industry as far back as the seventeenth century, elaborating further by saying that smugglers (who, legend has it, have many connections with the caverns) shipped the sand to Belgium, investing the proceeds in goods to be smuggled back to Britain.

Right Many of St Clement's carvings were the work of Joseph Golding, who opened the caves to the public in 1827.

St Clement's Caves

What is certain is that in the early years of the nineteenth century the caves had a reputation for harbouring a number of unsavoury characters, so much so that the entrance was sealed by the owner in about 1811. Rather promptly, it seems, the caves were then forgotten by the inhabitants of the town until their accidental 're-discovery' in 1825 by a Mr Scott, energetically wielding a pick-axe on the hillside above.

It was the inspired efforts of another local man, Joseph Golding, that eventually made the caves a tourist attraction. It was he who opened the caves for inspection, fully illuminated, in 1827, and who added so greatly to the original size of the place. With life-long enthu-

siasm, he dug new passages and chambers, the most striking being the large opening known as the Ballroom and a new exit gallery from the deepest part of the system.

Along with graffiti, carvings abound – many of them originating from the busy chisel of Golding. Many smaller inscriptions no doubt date from the last war, when the caves were put to use as a huge air-raid shelter. On average, 300 people sheltered here each night, and facilities included a school and a clinic. In its old days as a show cave, St Clement's was eerily silent, for its floors were covered in deep sand; with its enlistment as an air-raid shelter, the floors were concreted over.

Above A guide points out roof details in St Clement's Cave.

Young readers may like to make diary note to visit the caves on 14 October 2066, for then – on the 1000th anniversary of the Battle of Hastings – a capsule, sealed behind a wall plaque in 1966, will be opened. Inside is a copy of the mayor's speech made at the time of sealing, plus a copy of the local paper prepared in the 'style' of 1066. And if you missed it a century before, you will be able to hear the speech read all over again.

Scott's Grotto

Location: Hertfordshire
Scott's Road, Ware

Open:
By appointment with the warden,
telephone Ware 4131; a large torch is
required

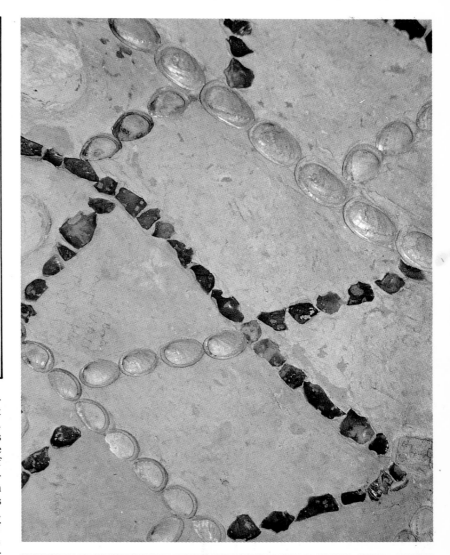

No one is quite sure why the eighteenth-
century poet and essayist John Scott
had his shell-lined grotto constructed.
But he may well have been motivated as
much by his pathological fear of disease
as by any desire to build a place of
beauty. So haunted by the fear of small-
pox was Scott that, although London
lay only 20 miles away, he visited it, it is
said, only once between 1740 and 1760.
Perhaps he felt an extra element of
security from the cool air of the grotto,
excavated from a chalk hillside. Certainly it helped to bring some of
London's society to his home, including Dr Johnson, who described the
grotto as a 'fairy palace'.

When the excavations were started is
not known, but they were completed by
1773. Not all the tunnels and chambers
were lined with shells and flints to the
extent of the Council Chamber, but it
has been estimated that Scott spent perhaps £10,000 on the construction.
When work on the grotto was abandoned, it consisted of seven chambers
connected by passages, with ventilation
and natural lighting vents at various

Right The magnificent, inlaid stone floor.

Scott's Grotto

points. The vestibule chamber, constructed outside, has not survived, but the rest is intact.

There are three doors at the front of the building, and the entrance is through the left-hand one. Steps lead down to the passage, with an air and light shaft entering above where it turns left. The first chamber, on the left, is the circular Refreshment Room, followed by – on the same side – the Committee Room, the only square one in the whole grotto. Branching off to the right is a short passage leading into yet another Committee Room: rather a lot of bureaucratic terminology for an away-from-it-all 'fairy palace'!

The main passage now curves to the right, enlarging into the Robing Room (some 70 feet in from the entrance and 34 feet below the surface), then reducing to two straight legs of passage leading to the Council Chamber. From this a few steps lead outside again. Between the exit and the entrance is a separate doorway, behind which a short straight stretch of passage leads into yet another, but solitary, chamber: the Consultation Room.

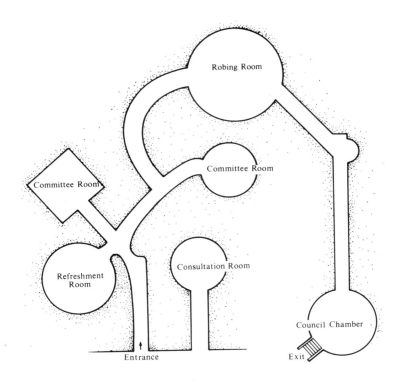

Below Stone houses of Scott's Grotto.

Hell-Fire Caves

Location: Buckinghamshire
On West Wycombe Hill, West
Wycombe, off the High Street

Open:
Daily, late February to late October,
1.00 pm to 6.00 pm,
early November to late February,
weekends, 12.00 to 4.00 pm

Stories abound about these caves but, without the evidence that no doubt still exists in private letters, notes and diaries written during their heyday, the stories remain speculation rather than hard historical fact. All the same, they are delicious.

During a series of harvest failures in the eighteenth century, Sir Francis Dashwood paid the agricultural workers of the parish to build a new road from West Wycombe to High Wycombe. True, the old road was in a terrible state, but his main intention was to provide the workers with employment. Work on the road continued from 1748 to 1752, and the large quantities of chalk it required were extracted from the hill behind an existing chalk quarry. Thus the caves came into being.

The caves, which are about a ¼ mile long, have a very unusual plan: at intervals opening up, or splitting and then rejoining, to form distinctive (and presumably symbolic) patterns. It has been suggested that their layout is linked with the Eleusinian Mysteries of ancient Greece, a mystical religious festival. A hidden cache of coins suggests that work on constructing the caves continued after the road had been completed.

The caves' main claim to fame

nowadays comes from their association with what is now known as the Hell-Fire Club, referred to at the time as the Knights of St Francis of Wycombe or the Monks of Medmenham. The Club was formed in 1746 by eminent men of the day. For nearly twenty years its activities were kept secret, meetings being held at Medmenham Abbey on the Thames. It is known a great deal of drink was involved and that ladies 'of a cheerful, lively disposition' had to be included, each disposing of 'a general hilarity'. The ladies considered themselves lawful wives of the brethren during their stay in monastic walls, every monk being scrupulous not to infringe upon the nuptial alliance of any other Brother.

It seems that a public row in 1763 led to a partial break-up of the club, and meetings were transferred to the caves at Wycombe. In various parts of the passages there are now tableaux with life-size representations of some of the members who escaped anonymity.

The cave entrance is half-way up West Wycombe Hill, and in the brick-lined approach tunnel are various historical documents about West Wycombe and the Hell-Fire Club. The first large opening is Whitehead's Chamber, named for the minor poet who was steward of the club. In the passage that follows, high on the left-hand wall, are the Roman numerals XXII. Though probably measurement marks, they may just possibly provide a clue to the whereabouts of suspected, but as yet undiscovered, secret passage.

The small maze of interconnecting passages which follows is the Labyrinth, from which a short stretch of passage connects with the Hall of Statues. Until 1972 the Hall, a chamber 40 feet wide and 50 feet high, was too dangerous to allow visitors access, but the roof has now been secured by an extensive stabilization process, with hundreds of steel reinforcing bolts and steel cables placed from the surface. The small by-pass tunnel running round this chamber was dug in 1954 by a former Yorkshire miner with the help of friends from High Wycombe.

After splitting into two to form an intriguing triangular shape, the passage becomes one again and descends sharply to the Treasure Chamber. Here were found seven coins, dating from 1720 to 1754, one bearing the scratched initials HV – presumably those of the foreman or mason in charge of the excavation.

Now, at a point 300 feet below the hillside surface, you come to a subterranean stream, inevitably named (in view of the caves' nefarious past) the River Styx. A bridge rather than a ferry leads one across the Styx to the Inner Temple. One can only guess at the sort of activities in which the Hell-Fire Club indulged here in the eighteenth century, but local stories are quite adamant: this was where the 'Wycombe Wenches left the last memories of their innocence'.

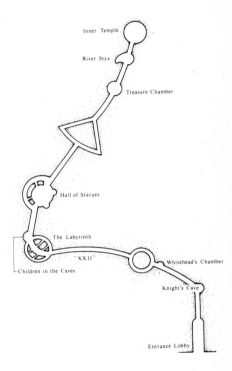

Chislehurst Caves

Location: Kent
Chislehurst

Open:
Daily, Easter to 30 September, 11.00
am to 5.00 pm; 1 October to Easter,
Sundays, 11.00 am to 5.00 pm; other
days by appointment, telephone
01-467 3264

The stories that abound about this very
complex series of passages in the village
of Chislehurst make it a popular tourist
spot, and inside it is easy enough to see
how the more inspired ones came
about. None of the passages is natural:
all were hewn out in the extraction of
chalk from the bed here, and, it is
claimed, the tunnels amount to some
miles in length. Just how old the caves
are is quite open to speculation. Some
evidence seems to suggest that the
Romans may have mined chalk here,
and they certainly needed large quantities of the mineral in and around
London. Possibly more imaginative are
the (admittedly rather tentative) suggestions that the oldest parts of the
caves were used by Druids, and sacri-
ficial altars have been 'identified'. Only
the less romantically inclined visitor will
actually give voice to the opinion that a
much more mundane explanation (of a
particular mining technique) is far more
likely.

Visitors see three main clusters of
passages on the tour of the labyrinth.
The outer series, at a place now known

Right An imaginative mural with a Roman
theme on the walls of Chislehurst Caves.

Chislehurst Caves

as the Church, includes what may have been a single denehole. Deneholes, which abound in certain areas of Kent and Essex, are distinctive subterranean excavations whose origins are really no clearer than those of Chislehurst Caves as a whole. A vertical shaft was dug from the surface down to the chalk, where side chambers and passages were opened out into the body of the rock, often in a clover-leaf formation. These remarkably stable excavations probably, though by no means certainly, date to sometime between the twelfth and fifteenth centuries, and their most likely purpose was to supply chalk for agricultural use. While we can apply the full force of modern science to dating samples of rock brought back from the face of the moon, it seems we can be baffled over the age of large, man-made structures on London's doorsteps.

As fascinating and mysterious as the cave's origins are, its later, recorded, history is just as interesting. During the Civil War in the 1640s, a number of Royalist mansions in Chislehurst had secret bolt holes into the caves, used to hide valuables or even as a means of escape. A century and a half later, when

Britain was threatened by a Napoleonic invasion, flints were knapped (shaped) in the caves for flintlock rifles. In the First World War, the caves' high security led to their use for storing large quantities of munitions, and light railway tracks were laid to shift stores.

At the start of the Blitz in the Second World War, the caves were once again called into use for protection – not of ammunition this time but of people. The great complex of passages, deep out of harm's way, became the country's largest air-raid shelter, as many as 15,000 people taking refuge there each night. At first, illumination was by candlelight and the shelterers slept on the hard floors. As the enterprise grew electric lights and bunks were installed, adults being charged one old penny a night. With a nightly population equivalent to that of Cirencester, for instance, supervision had to be carefully handled, and 'Cave Captains' were appointed, each responsible for 200 people. And from basic beginnings, virtually a small town grew, with shops, churches, a cinema, dance hall, library, and so on. Two doctors were on call, but the health of those who stayed in the

caves for any length of time was said to be excellent. Many sufferers from asthma, bronchitis or a particular form of rheumatism reported their health improved, and after the war several pharmaceutical companies tried to trace the source of this curative property, but without success.

Ghost stories are thick on the ground in Chislehurst Caves, the focus of many being the pool in which, legend has it, a woman was murdered by drowning 200 years ago. Several men have steeled themselves to a night alone in the passages, one lasting the night, the other only twenty minutes. Neither, reports the owner, would repeat the experience for any amount of money. And the cave's foreman, who after nearly three decades there can walk through the passages without a light, has on occasions, in one particular area, felt a hand lightly grasp his arm.

Right The Roman Well in Chislehurst Caves.

Below The Map Room in Chislehurst Caves.

Royston Cave

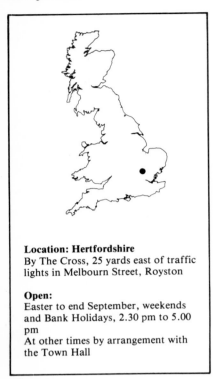

Location: Hertfordshire
By The Cross, 25 yards east of traffic
lights in Melbourn Street, Royston

Open:
Easter to end September, weekends
and Bank Holidays, 2.30 pm to 5.00
pm
At other times by arrangement with
the Town Hall

PLAN

N

ELEVATION

Steps

Passage

Cave

Melbourn Street

Street level

Chalk

Right Close-up studies of some of the
carvings in Royston Cave.

Set into the pavement outside a bank in
the little town of Royston is a small
metal grille. Beneath it lies a man-made
hole, the origin of which continues to
keep the experts guessing: Royston
Cave.

In the summer of 1742, while working on the shelter used by the women of
the Butter Market, workmen uncovered
a millstone. Lifting it, they found a
cavity, and down this was despatched
first a boy, then a slender man. They
reported a bell-shaped chamber carved
from the solid chalk, half filled with
earth and debris. After this was cleared
the walls were found to be almost
covered in primitive carvings. When
discovered, the cave was entered via a
shaft, but in 1790 an access tunnel was
dug for 72 feet to enter the base of the
cave at the only point where there were
no carvings to be damaged. It is through
this tunnel that you now reach Royston
Cave.

The chamber is roughly 20 feet in
diameter, and rises to a maximum
height of about 30 feet. There are two
(sealed) shafts entering the ceiling as
well as the central hole now capped by
the metal grille. Half of the millstone
which originally capped the chamber
forms the bottom-most of the entrance
steps, the other half is on display.

When the cave was discovered there
was evidence that the carved figures had
originally been coloured. But this is
now no longer discernible except just
below one of the shafts where what
appears to be red brick is, in fact,
painted chalk.

The subjects of the carvings are in
some cases quite obvious, in others
open to differing interpretations. The
majority of the figures represent saints,
either marked by crosses or bearing
some easily identifiable symbol: St
Lawrence, martyred by being roasted
alive on a grid iron; St Katherine,
martyred on a spiked wheel; and St
Christopher, seen with staff in hand
and the child Jesus on his shoulder. The
figures at the base of each crucifixion
are most probably those of Mary and
John, and the largest of the various
niches cut into the wall may well be a
representation of the Holy Sepulchre,
with the damaged figure showing Christ
awaiting resurrection.

But one turns, inevitably, to the
question: who made Royston Cave, and
why? And when were the carvings
made, and by whom? Were such a site
to be found these days, first priority
would be given to painstaking archaeological excavation. This was not the case
in 1742, unfortunately, so what clues
there may have been have been removed
or eradicated. However, there are

pointers. The carvings, most believe,
date from about the 13th century. The
cave is situated very close to the
junction of the prehistoric, and later
Romanized, Icknield Way and Ermine
Street, the Roman road from London
to York — and in many ages crossroads
have been regarded as sacred. Some of
the most recent research links the cave
with the Knights Templar, the powerful
order of military knights outlawed by
the Pope in 1312. There are numerous
local connections with the Templars,
and good reason to suppose that they
would have needed to construct some
Templar chapel in Royston. They are
known to have attended Royston
market regularly, and were required
(when not on active duty) to make
frequent daily devotions. But no one
can be certain, the cave keeps its secret.

Grimes Graves

Location: Norfolk
5 miles north-west of Thetford, off
the B1108

Open:
Daily, 1 April to 30 September, 9.30
am to 6.30 pm
15 to 31 March and 1 to 15 October,
weekdays, 9.30 am to 6.30 pm;
Sundays, 2.00 pm to 6.30 pm
16 October to 14 March, weekdays,
9.30 am to 4.00 pm; Sundays, 2.00
pm to 4.00 pm

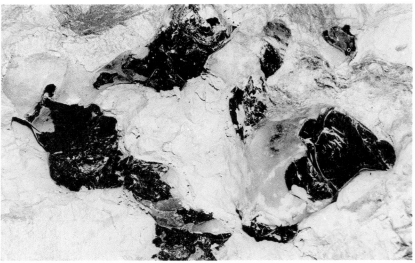

First World War battlefields, where the
guns and mortars have reduced pastures
to the pock-marked face of hell – that is
the image that most readily springs to
mind at Grimes Graves. And the name,
of course, serves only to reinforce this
impression. In fact, the craters – well
over 300 of them – are reminders of a
peaceable industry carried on 4000
years ago. As for the name: Grim is
another name for Woden, chief of the
Anglo-Saxon gods, and Graves merely
a term for holes or hollows.

Over the centuries many theories
were put forward to explain the strange
cluster of hollows at Grimes Graves.
The truth emerged in 1870 when, having
excavated one of the hollows for three
years, a Canon Greenwell was able to
show that they were flint mines dating
back to the Neolithic period.

In the area here of about 34 acres
cared for and administered now by the
Historic Buildings and Monuments
Commission, there are approaching 800
shafts and shallow pits, many filled
flush to the surface, leaving no tell-tale
hollow. Many have been dug out by
archaeologists since the pioneering days
of Canon Greenwell, and many of the
questions which spring to mind about
such an ancient industrial site can now
be answered.

The best place to appreciate these

remarkable workings, and the efforts
put into making them, is of course
inside. Just one, Pit 1, can be descended
to its working floor, about 30 feet deep.
The largest shafts are between 13 feet
and 26 feet in diameter at the top and up
to 46 feet deep.

At the foot of Pit 1 the entrances to
the seven low galleries can be seen, varying from 5 to 3 feet in height. Here the
Neolithic miners crouched as they followed the bed of flint. What drove them
to work at such a depth was the high
quality of the flint, now known as floor-
stone, which ran at this level.

Without the security provided by pit-
props, the miners evolved a method of
safely extracting the maximum amount
of flint from a given area. The pits were

dug close together, and belled out as
they got deeper. Often the galleries dug
from the bottom were driven until they
intersected with others from adjoining
shafts; then sections of the gallery walls
were mined away, leaving a sufficient
number of pillars to support the roof.
Fully-worked galleries were frequently
used to dump chalk spoil from fresh
galleries, which saved much effort.

The chief mining implement was the
L-shaped antler of the red deer, and it is
estimated that upwards of 150 were
required to dig each shaft. While daylight was sufficient illumination for
this, it is uncertain whether artificial
lighting was used in the dark galleries. A
number of chalk cups have been found
which could have been fat lamps.

Also unknown is exactly how the miners climbed the shafts, both to get to the workface and to raise the flint and chalk spoil. Obvious solutions are a tree trunk with most of the branches lopped off (clumsy but effective), a leather thong and wooden-rung 'rope' ladder, or – according to the most recent thinking – straightforward wooden ladders. But no finds have yet come to light that can settle this question.

Naturally enough, we tend to be fascinated with the manpower involved in such human efforts: how many men did it take to raise Stonehenge or the pyramids? The archaeologists who re-excavated Canon Greenwell's pit estimated that it would take a team of twenty men between eighty and a hundred days to dig a typical shaft 40 feet deep, though only half a dozen or so would be able to work in the confined area at any one time. Four galleries would have taken about six weeks to dig, yielding about 36 tons of flint but more than 800 tons of spoil. This last figure shows the common sense of using worked-out galleries for dumping spoil.

And the flint which was so laboriously mined and brought to the surface? Axes and knives accounted for the majority of tools fashioned at Grimes Graves, though others were made as well. About 4000 flakes of flint had to be struck off a flintstone to make just one axe, and 6-foot-thick deposits of flakes have been uncovered on the surface.

Above The surface hollows of infilled shafts of Grimes Graves.

Top Left One of the excavated tunnels of Grimes Graves, originally blocked with chalk rubble after Neolithic Man had finished his removal of flint. A modern prop now supports the roof.

Bottom Left Chert nodules, or flint, still embedded in the walls of one of the excavated shafts of Grimes Graves. This is what Neolithic man was mining for.

WALES & THE BORDERS

Many of the villages and towns of South Wales are based around some of the greatest coal mines of Britain. From the richly bituminous deposits of the east, through the steam coals of the central valleys (once so important a source of supply for steam ships) to the deep, rich and hard anthracite seams of the western fields, coal has been king.

At Blaenavon, the Big Pit Mining Museum gives a fascinating insight into how coal is brought from the earth, and the conditions in which miners worked, from the days of pick and shovel to the most modern machinery. Big Pit enables one to understand – at least a little – what it really means to work underground for a living.

While we look to the valleys in South Wales for underground exploitation, in North Wales it is the mountains that hold our attention. Beneath some of the peaks, faces and ridges of Snowdonia above Blaenau Ffestiniog are great corridors and caverns hewn in the chasing of the region's rich slate beds. Two of the great slate mines are open to the public: Gloddfa Ganol Slate Mine and the quite separate Llechwedd Slate Caverns. The tour of the first is on foot, while Llechwedd offers a choice of transport: a level tramway or a specially built carriage plunging steeply down to the deeper workings.

A geographical and thematic counterpoint to the great coalfields of South Wales is provided in the north by Dinorwig Power Station. Both are concerned with energy, but at Dinorwig – a massive engineering undertaking almost entirely underground – the pumped-storage layout enables surplus electricity generated at night to be stored for release in the day.

A much older story is told at the Dolaucothi Gold Mines in Dyfed. This is the only place in Britain where the Romans are definitely known to have mined gold.

In the border county of Shropshire, serendipity played a large part in the discovery of a spring of natural bitumen 200 years ago. Much was extracted, but black glistening pools still accumulate slowly.

What of the natural subterranean attractions in the region? Cavers are slowly exploring and opening up a limited number of systems in North Wales, but it is in the south – almost entirely within the 40-mile-long Brecon Beacons National Park – that the really great Welsh caves (or *ogofs*) are found. As in Mendip, there are relatively few deep shafts. But there is ample compensation in the vigorous underground rivers and streamways where the clamour and roar of the torrent, especially in high water conditions, echoes down the dark-walled passages.

Here measurements in miles rather than feet are common: the giant system of Ogof Ffynnon Ddu, with some 24 miles of passage; Ogof Agen Allwedd, totalling nearly 16 miles; and Dan-yr-Ogof, with 9½ miles. This last system reveals some of its secrets to the non-caver, too, in the show cave in its early sections. Close at hand is Cathedral Cave, without an active streamway but with a huge and majestic main passage.

Big Pit Mining Museum

Location: Gwent
Off B4248 between Blaenavon and Brynmawr, 3½ miles from Blaenavon

Open:
Daily, 1 April to 31 October, 10.00 am to 5.00 pm (last tour starts 3.30 pm); tours do not operate on Mondays. At other times by appointment

A typical Welsh coal mine, situated on the eastern rim of the great South Wales coalfield, Big Pit seemed likely to join the dust and anonymity of other worked-out pits when it closed in 1980. Thanks to the rising interest in industrial history, and the growing number of people keen to see at first hand just how miners have won the precious black mineral over the years, it was preserved. The establishment of the Big Pit Museum Trust — supported by a number of bodies, from Gwent County Council to the European Regional Development Fund — ensured that it would live on to provide a fascinating insight into the workings of a pit.

The year 1789 was a key one for the region, for it was then that the first iron-smelting furnace was built at Blaenavon. Bit Pit — created in 1880 by enlarging the older Kearsley's Pit — yielded much coal for this industry, and in its earlier days even ironstone from the thin beds along the northern edge of the coalfield. Before long the coal was also being used to power factories, trains and ships all round the world,

and by 1910 more than half the coal from South Wales was being exported. The 1930s brought great unemployment in South Wales, although rearmament in the late 1930s and then nationalization in 1947 brought the miners their jobs back. But at Big Pit it was the nature of the coal deposit rather than economics that eventually dictated its closure. With the miners lying flat-out to extract coal from the last seam, the roof height became too low for approved supports to be used and the last ton of 300 million-year-old coal rattled its way to the surface.

Before the descent into the workings you are issued with a helmet, electric lamp and self-rescuer, and must leave in safe keeping any contraband — not French brandy and perfume but matches, lighters, and anything else that constitutes a fire risk in the mine.

Below The pit head and other surface structures at Big Pit.

Big Pit Mining Museum

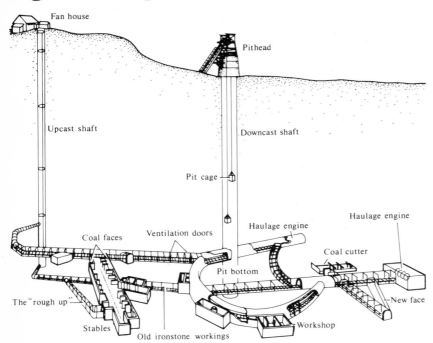

Fan house
Pithead
Upcast shaft
Downcast shaft
Pit cage
Haulage engine
Haulage engine
Coal faces
Ventilation doors
Coal cutter
Pit bottom
The "rough up"
New face
Stables
Workshop
Old ironstone workings

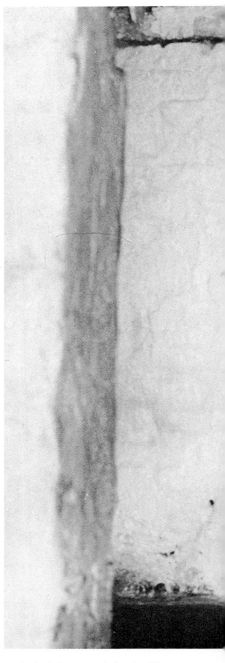

The journey down the 300-foot deep brick- and stone-lined shaft in the cage takes just over a minute, the descent being matched by the ascent of the counter-balanced cage. From the pit bottom, your eyes slowly becoming accustomed to the wide beam of light cast by your lamp, the route starts along a short stretch of passage, passing a steam pump of the type used before the pit was electrified in 1910. The roadway ahead originally continued to two of the old workings. Throughout its history, nine different seams have been worked in Big Pit.

A curving roadway leads to the long stretch along Black Vein, which is followed eastwards at first to a large haulage engine (one of a number in the pit) and the pit's latest display, an example of a long-wall working with a mechanical under-cutter, typical of the type used in Big Pit in about 1940. In long-wall working, as the name implies, a long stretch of the seam was worked in one pass, the under-cutting leaving a cavity for the coal to fall into. In this direction too, incidentally, lies the mine's 'second way out'.

Westward now, branching off left into a roadway which curves back almost to the pit bottom again. Along here, the example of timbering clearly shows the use of the 'Welsh notch'. Because the rock strata dip in this area, the props supporting the roof beams must also be able to cope with possible sideways pressure. This is achieved by shaping the top of the prop to fit into a notch cut into the end of the roof beam. Grim reminders of the ever-constant danger in mining are the stretcher and trambulance on which casualties were evacuated to the surface. The fitting-shop on the left was used by the pit-bottom fitter; each 'district' in the mine also had its own fitter, two later on as mechanization increased.

A sharp left turn at the end of this curving roadway, away from the pit bottom again, takes you along another with three ventilation doors. Big Pit is ventilated by drawing the air down the Big Pit shaft (the downcast), through all the various roadways and workings, then up the separate Coity Pits shaft (the upcast) to the huge ventilating fan on the surface. The ventilating doors prevent the air taking the shortest route to the upcast shaft – a simple but effective method of ventilation that has been used since the 18th century. A government inspector visiting the Blaenavon area in 1842 reported that 363 children were working in the coal and iron mines here, many of whom began their working lives as young as ten years old, operating such air-doors. The doors are arranged in threes so that trains of trams (known as journeys) could pass through the air-locks without disrupting the ventilation pattern.

Further along are examples of small and large haulage engines; the smaller is typical of those made by the Blaenavon Company in its own workshops. The overhead haulage cable, to which the trams were attached for the journey to and from the pit bottom, is in the form of an endless loop.

Just inside the entrance to the stables are two displays of coal-mining by hand. On the right is the pillar and stall method, used widely in South Wales until the last part of the 19th century. A stall had a narrow entrance to support

the roof, and then widened to about 27 feet at the coal face, and was worked by a collier and his boy. On average they mined 5 tons of coal each day, and before child labour was abolished children were often used to drag the coal to the trams on sledges. On the left is a representation of a hand-cut longwall face, used from the 1890s to the 1930s.

In the stables ahead were kept the horses used for haulage on the main roadways before the introduction of steam haulage. In 1930 Big Pit had seventy-one hauliers and about the same number of horses, and the last two horses – Victor and Impact – only finished working here in 1974.

From the stables the Rough Up – a passage driven for ventilation purposes, and with ironstone visible in the roof – leads to the Mine Slope. This, the oldest part of the mine shown on the tour, was originally driven from an entrance near the river to work the ironstone, and

Above The sunless accommodation for the dozens of horses used for haulage before the introduction of steam power.

Big Pit Mining Museum

runs at a gradient of 1 in 12 for 1000 yards. This roadway takes us back to the shaft, past the New Cut heading (directly beneath the air-doors seen earlier) and a variety of roof supports, and the old ironstone workings.

Back on the surface there is still plenty to see: the winding-engine with its vital automatic controlling system; the blacksmith's shop, with the forges away from daylight to assist the testing of colour of heating iron; and the pit-head baths. Opened in 1939, these greatly improved miners' lives. No longer did they have to return home in their pit clothes for a bath in a tin tub before the cottage fire, and the prospect of being able to don dry, aired working clothes at the start of each shift must have taken at least some of the edge off that long drop underground to a hard day's work at the coalface.

Top Left Mining techniques demonstrated in Big Pit.

Below The River Arch in Big Pit.

Dan-yr-Ogof Cave Complex

Location: Powys
On A4067 in Brecon Beacons
National Park, about 18 miles from
Swansea on the way to Sennybridge

Open:
Daily, Easter to 1 November, 10.00
am to 5.00 pm
For details of winter opening,
telephone Abercrave 730284 or
730693

Above The exquisite straw formations in
Flabbergasm Chasm, Dan-yr-Ogof.

Not far from the southern edge of the
Brecon Beacons National Park, whose
soft contours give lie to the fact that in
places they rise to nearly 3000 feet, is
this grouping of two fine show caves
plus (the most recent addition) a rich
bone cave. From the assembly-point
just past the ticket office, which lies
over 800 feet above sea level, you have a
fine view of the Upper Tawe valley.
Up the valley are the slopes of Fan
Gihyrych, one of the Beacons of
Brecon. Closer is the river Tawe, and
the delta of the river Haffes, while due
south is the Gothically-styled 'castle' of
Craig-y-Nos, once home of the famous
prima donna, Madame Adelina Patti.

Dan-yr-Ogof

The name means, literally, 'below the
cave' – an odd title to give any cave,
particularly one of the country's largest
show caves. But it was christened thus
in the days when much more attention
was being given to the archaeologically
important Ogof-yr-Esgyrn, which lies
about 100 feet higher.

The story of the opening up of Dan-
yr-Ogof is one of dogged perseverance
over the years. In June 1912, two
brothers from Tymawr, Abercrave –
Ashwell and Jeff Morgan – first
entered the river cave. It was Jeff who

found and squeezed through a small
hole in one of the banks, his cries of joy
echoing back to his brother as he came
upon a large and well decorated passage
high above the river. The two pushed
their way forward along the sandy-
floored passage until a deep pool barred
the way.

Over the following days, accompanied by another brother, Edwin, and
their gamekeeper, using a small Welsh
coracle to negotiate the water, and with
candles as their only form of lighting,
they pushed their way ever deeper into
the mountain. On one trip, on a shelf by
the magnificent formation known as
the Alabaster Pillar, they left their
thermos flask and a bottle containing a
note of their names. The remains are
still there to this day. Their progress was
halted when they came to a lake, placing
enough, but with the loud roar of a
waterfall coming from the darkness
beyond. A larger coracle was obtained,
and Ashwell Morgan set out across the
water while his brother paid out a rope
attached to the craft. His courage was
considerable, for his frail craft might
well have been swamped – or even

sucked down into some unknown water-
course if the roar had come from water
pouring *out* of the lake. He found not
one but three lakes, and into the far end
of the final one poured an intimidating
and vigorous waterfall.

There exploration halted until
twenty-five years later. The next phase,
when experienced cavers from Yorkshire and Mendip took over, started in
1937. Their first venture was plagued by
trouble with the inflatable boat and a
strong current, and it should go on
record that the first to pass the waterfall
was, in fact, Ashwell Morgan, his old
enthusiasm rekindled. With another
man and a plucky female doctor (who
swam the whole way back), he crossed
another 150 feet of deep water and
yet another waterfall, stopping only at
the daunting sight of a fourth lake
stretching into the distance.

The caving teams doubled the known
length of Dan-yr-Ogof that year, their

Dan-yr-Ogof Cave Complex

explorations ending in an extremely tight crawl. Meanwhile, the Morgan brothers were working at opening the cave for the public. A new entrance was made above the river entrance, paths were concreted, and electric lighting installed, power coming from a turbine generator at the Llynfell Falls which can be seen on the walk up to the cave.

Opened in 1939, the show cave had a short-lived, if successful, debut. Closed at the start of the Second World War, Dan-yr-Ogof did not lie redundant; it was used as a government store for large quantities of explosives. Various difficulties then prevented the cave from being opened for nearly twenty years, though occasional visits were made by cavers, and a few extensions made to the known system.

After the re-opening to both the public and cavers in 1964, the South Wales Caving Club mounted trip after trip in an effort to pass the Endless Crawl – the tight section marking the end of exploration those many years before. All were fruitless until, in 1966, Eileen Davies passed the squeeze. As tight as the constriction was, the psychological barrier had been just as great. With the knowledge that one could turn round on the other side, larger cavers found that they could follow! Today the notorious squeeze is passed by very few, for a by-pass has been discovered. Since then, employing at times what can only be described as siege tactics with six-day long underground camps, cavers have extended the cave until it ranks among the longest

in Britain, with over 9 miles of passage.

While the discoveries of the later cavers remain the domain of the very experienced caver, those of the Morgan brothers, as far as the first lake, now lie before you.

Just inside the short mined section by which you enter Dan-yr-Ogof is a door, clearly marked, behind which is the continuous roar of the emerging river. This is where the Morgan brothers first made their way into the cave. Immediately beyond, a bridge spans a deep pool – your easy way over the first obstacle met by the brothers, and crossed in that tiny coracle. And here is one of the first memorable formations: the Frozen Waterfall, a tumbling cascade of flowstone.

Rounding a corner you come to the

Belfry, where stalactite 'ropes' dangle from the bell-like boulders lodged in the roof. The white sheets plastered overhead are moon milk — a strange, soft calcite deposit, quite unlike any other cave formation. High on the left, where the passage begins to widen, is what remains of the Pencil Column: an enchanting formation that has proved irresistible to some vandal. Below, close to the floor, is a fine specimen of a coral colony — a reminder that the cold rock about you was formed at the bottom of a warm tropical sea over 300 million years ago.

Immediately below a group of short, white straw stalactites known as the Pin Cushion is a junction — the Parting of the Ways. The route inwards goes up a flight of steps to higher passages. At the top are the remains of the flask and bottle left by the original explorers, but it is the Alabaster Pillar which most catches the attention. A white column, 6 feet long, joins a low ceiling to the floor, its spreading base sloping into the still waters of a small reflecting pool. An ascent round to the other side of the Pillar reveals the Flitch of Bacon, an excellent curtain which is back-lit to reveal the bands of different colours.

The next stretch of passage offers a variety of formations, and an opening known as the Window, through which can be seen the lower passage, the outward route. But the most spectacular sight is yet to come, in Cauldron Chamber. From the 40-foot-high dappled roof hangs one of Britain's finest calcite curtains, 18 feet long. Again, back-lighting has been effectively used to show up the gorgeous translucence.

From the static glory of the Curtain, a descent takes you back to the vigour of the streamway in Bridge Chamber and the first major obstacle which faced the Morgans — the First Lake. This is now spanned by a bridge (not recommended for the elderly or infirm because of the lowness of the roof for several yards), occasionally closed in very high waters. This bridge leads to Jubilee Passage and the end of the journey inwards, where the light sand on the floor by the start of the long Second Lake contrasts starkly with the black limestone walls. You are about ¼ mile from the entrance, and above are 300 feet of rock. Ever-present here is the thunder of the river which carved this system.

And it is on the return journey that the often 'soft' sculpting of the walls, so typical of a number of Welsh caves, is in evidence in Western Passage. This is the newer (in geological terms) passage which you can follow from Bridge Chamber through to the Parting of the Ways, where the inward route is rejoined.

Cathedral Cave

Tunnel Cave, cavers still call it, and in view of what will be revealed this seems particularly inappropriate. But, like many caves, it was named after some peculiarity or notable feature of its entrance, and for many years this was simply a tunnel-like cave above Dan-yr-Ogof, 150 feet long and ending in a boulder choke, and often flooded to the roof. As you might imagine, early enthusiasm for this hole was not high.

However, the insignificant hole did have something going for it, something very precious to the caver's heart: a mighty draught. A strong draught is the sign of a great cave system breathing, and has lured many a caver to persist just that bit longer with a stubborn dig. The South Wales Caving Club did persist, first clearing a way through the unstable boulder choke and then, late in 1953, through the boulders beyond. A small hole was opened, and a great deal of darkness seemed to lie beyond. Not

Left Formations in Dan-yr-Ogof.

Below Backlighting in Dan-yr-Ogof reveals the translucent beauty of the curtain called the Flitch of Bacon.

Dan-yr-Ogof Cave Complex

surprisingly, for they had discovered a way through into what is now possibly the largest chamber illuminated by floodlights of any British show cave.

Many show caves lead to 'wild' sections which only the competent caver can safely negotiate, but Cathedral Cave is unusual in offering the caver a choice of two routes to the non-tourist parts: via the show cave section itself, or through a top entrance that was discovered from the inside as the original explorers pushed higher and higher. As you walk up to the entrance of the show cave, glance over the railings to the right. You will see the original entrance, the small and flood-prone passage that first gave the system the prosaic name of Tunnel Cave. The artificial entrance you now make your way along was opened in 1971, and leads straight to the full majesty of the main hall. Look right again here; the steel grill bars the way through which cavers first made their way in 1953.

The large size of the passage up which you are making your way is impressive, yet it will become greater, right up to the end of the show cave. As a consequence, the formations along the way are placed in a quite different perspective from those in Dan-yr-Ogof. In this early section, one remarkable feature is the Carrot – a formation which started life as a thin straw, then later burgeoned into the heavy tapered shape for which the name is so apt, and which seems in imminent danger of becoming just too large for the remaining slender portion of straw stalactite to support.

At the first bend, on the right-hand side, is the 'Museum'. This is a collection of the formations that had to be moved when the show cave route was being laid. A little further on is the water-filled passage known as Flood Rising. In 1954, after an exceptionally dry period, a caver was able to follow a small airspace along here for 100 feet. Since then, divers have probed ever deeper, and the challenge still stands.

Large stalagmites and columns occur more and more frequently past this point, and on the left of the path are numerous examples of the various phenomena created by falling water, such as drip pockets and cave pearls. The spread of wall with the name of Organ Pipes owes its intriguing configuration not to the depositing of calcite, but to the slow erosion of the limestone.

The passage narrows slightly, taking the profile of a gorge, then blossoms

out into the high and wide Dome of St Paul's. On the left is a lake, fed from a high waterfall, and the water enters at a point which marks the level of a number of passages from which hang arrays of stalactites. The tour ends at a section whose shape is perfectly summed up in its name, Keyhole Passage, and this profile is encountered in much of the rest of the cave, the total length of which is 7000 feet.

From the caver's point of view, it is satisfying that tableaux are displayed in Cathedral Cave depicting the various methods of exploration employed over the years. The first exhibit shows the Morgan brothers and their equipment in 1912, when they first pushed back the barrier of the unknown in Dan-yr-

Ogof, with candles, oil lamps and a coracle. In the Dome of St Paul's are three exhibits showing various means of descent and ascent used by the modern caver. The long stretch of jointed aluminium poles is known as a maypole, a simple if sometimes precarious means of raising a ladder to an unexplored entrance at a higher level. Another display shows the two principal means of ascending and descending cave pitches: ladders and single rope techniques (see page 28). Finally, there is a tableau representing a typical underground camp on a caving expedition. The tent is not quite as anomalous as might be first thought, for it provides a warm and psychologically reassuring place in which to nestle.

very important for one of the key dating methods. More than 10,000 finds were made, the majority fragments of human and animal bones. The main burial-point was the sand pit beneath the 'X' painted on the wall, where the bones of at least a dozen men and women were found, dating back to the Roman era.

As well as plenty of evidence of occupation in Roman times, Ogof-yr-Esgyrn yielded a smaller number of finds indicating occupation in the Bronze Age. Among them were one of the very few dirks or rapiers from that time ever found in Wales, and, rarest of all, a bone weaving-comb made by cutting teeth into an ox rib.

The main excavations were completed in 1950, but a further, limited, search was made between the boulders on the floor in 1976, and in the following year a large fragment of the skull of a seven-year-old was found embedded below the stalagmite boss in the centre of the chamber.

A number of displays have been mounted in the cave to illustrate archaeological excavation in such caves. The first shows the instruments used in surveying a typical dig — a surveying pole and dumpy level on a tripod. Both drawing and photographic techniques are illustrated, as well as a large metal grid with a mesh of string at 4-inch spacing; this is placed over the area being dug to give a precise reference-point for each find. In another area is a sieve used for separating fragments from excavated soil. The example shown has rather a coarse mesh; modern digs often have additional sieves with a fine enough mesh to hold back seeds and even pollen grains.

Another scene shows a number of human skeletons as they might be exposed after excavation. Recording the exact position of each bone is very important, for this may well give clues about when the skeleton was buried, and possibly the cause of death too.

Life-size models of four of the animals that used caves for habitation include the enormous Cave Bear, about 9 feet high when standing upright, and the Smilodon, a voracious sabre-toothed cat. There are also life-size models of a Bronze Age grandmother and child (though bear in mind that they would probably have lived outside the cave, using it only for shelter and storage) and a Roman soldier, based on descriptions of a soldier serving with the Second Augustan Legion.

Ogof-yr-Esgyrn (Bone Cave)

Situated about 140 feet above Dan-yr-Ogof, this cave consists of an almost rectangular chamber measuring roughly 60 feet by 36 feet. Tiny compared with its nearby brethren, it has nevertheless proved a rich source of archaeological discoveries.

Originally called simply Yr Ogof (The Cave), it took the more specific name when human and animal bones were discovered there, chiefly to save confusion with Dan-yr-Ogof. The first known excavations were made by students, under expert guidance, in 1923; they found a number of personal ornaments plus various fragments of human bone. The main period of excavation lasted from 1938 to 1950,

Above Bone Cave, where the remains of about 12 people were found, believed to date back to the Roman Era.

Left The Cathedral Cave.

and presented a number of problems. The floor was liberally covered in boulders and blocks of stone from the roof, all of which had to be shifted with great care to preserve any remains underneath, and the crust of stalagmite flow over the floor had to be pierced.

Before a careful examination of the floor deposits could be made, the whole chamber was divided into 1-yard squares (British archaeologists had not been metricated then), so that the precise position, and depth, of each find could be recorded. Knowing the exact strata from which finds come is

Dolaucothi Gold Mines

Location: Dyfed
Off the A482, ¾ mile south-east of
Pumsaint

Open:
Daily, 18 June to 11 September, 10.30
am to 5.00 pm

Gold! Throughout history, the word
has evoked a particular response quite
out of proportion to its monetary value,
related rather to its sheer immutability.
While paper money and assorted coin-
age come and go, gold persists; so does
its attraction.

Dolaucothi is the only place in Britain
where the Roman are definitely known
to have mined gold, and many of their
workings here – plus those of other,
later times – remain for us to see. The
native population may also have mined
gold here too, on a small scale, but any
positive evidence has disappeared.
After the Romans arrived at Dolaucothi
in about AD75, and established a fort
where the village of Pumpsaint now
lies, extensive development of the site
soon got under way.

A copious supply of water was vital
to Roman methods of mining. A technique known as hushing was used, in
which a controlled flood of water
washed over the open-cast working
area. This first removed the top covering, then mined debris. The water was
brought to the site by a series of
aqueducts and stored in reservoirs
above the working area.

Whether any mining was done here
from the time of the Romans to the
nineteenth century is not known, but
the discovery in 1844 of a tiny particle of
the precious metal at Dolaucothi
renewed interest. The area was mined
intermittently from 1872 to 1938, but
lack of funds or poor yields of gold led
to each enterprise closing.

There are a number of recommended
tours around the many remains and
workings, but we shall concentrate here
on the guided underground tour which
takes in various underground workings.
The most noticeable feature of the
landscape is the Ogofau Pit, a Roman
open-cast excavation measuring 500
feet by 330, and about 80 feet deep. The

Right Long Adit Portal, Dolaucothi Gold
Mines.

Below The entrance to Pumpsaint Gold
Mine.

Dolaucothi Gold Mines

Above Mitchell Adit: a massive timber roof-support in worked-out gold stopes.

Left Mitchell Adit: a lamp beam catches the glint of a gold vein.

original floor level – which would have been much rougher when the Romans followed the richer veins – lies a further 15 to 30 feet down, buried beneath waste rock which was spread to provide a flat foundation for twentieth-century mine buildings. At the end of the pit closest to the car park, the old miners went underground, working the upper part of the Roman lode to depths of over 100 feet.

The Romans removed nearly one million tons of rock from the Ogofau Pit. You realize the scarcity of gold when you consider that the total volume of gold recovered from all that rock would probably occupy a cube with sides between 14 and 18 inches long. But then, on a global scale, it has been estimated that all the gold *ever* produced could be contained in a cube with sides the length of a cricket pitch.

Take the path which starts by the side of the Mine Office beneath a small adit and goes up the north-eastern side of the pit. Halfway up the steep rock face on the left is the large entrance to the Middle Adit, above which, on the left, a fold in the rock is clearly visible. Formed from deposits in the sea bottom 440 millions years ago, the rocks were later subjected to tremendous pressures, resulting in folds such as this.

The rear wall of this cavernous adit lies about 70 feet in from the entrance, but take a look at the right-hand wall about halfway in. Here is another indication of the restless nature of the earth's crust, for the strata here were twisted some 390 million years ago from their original horizontal plane to the present near vertical. They show well-preserved ripple marks.

Gold-bearing shales can be seen on the rear wall, with their distinctive yellow-white weathering, and it was these that the Romans were following when they excavated the adit. They appreciated the good yield of gold which could be had from them, but they relied heavily on weathering having broken down the shales, releasing the minute grains of gold. Such weathering of course diminished the deeper they dug, dictating their final mining limits. The small adit which runs for about 80 feet from the rear wall was probably a trial dig made in the last century.

Turning left after the exit from Middle Adit, the route takes you round to the gaping mouth of Mitchell Adit, the longest underground part of the tour. The first part was, it seems, worked before the 19th century, but it takes its name from James Mitchell, who was responsible at the beginning of this century for excavating it to its present extent. It runs fairly straight throughout its length of roughly 260 feet, with various workings leading off; there are thin but rich gold veins at the

Fiveways Junction, some 150 feet in. The entrance to the stopes (the holes left after the extraction of the ore) runs from the crossing passage at the end of the adit, and it was also from here that Mitchell sank an internal shaft to meet the Long Adit driven from the Ogofau Pit floor.

The stopes emerge at the surface in an open-cast working, probably dating back to the Romans, where two main veins of quartz have been mined. The walls of one are particularly roughly worked where the vein so firmly adhered to the shales that it had to be chiselled or pick-axed away.

As well as the underground tour at Dolaucothi (which is under the care of the National Trust), there are several surface tours round the complex of workings, both ancient and relatively modern. All start at the Mine Field Centre Office which belongs to the Department of Mineral Exploitation at University College, Cardiff.

A final point. As you wander round Dolaucothi you might well, and not unnaturally, ponder on whether there is still gold in these 'ere hills. The answer is yes. But only the future can tell whether the gold-bearing veins will ever again be followed into the depths. In view of the enormous efforts made here over two millenia to mine gold, it is interesting to note that perhaps a third was actually thrown away with the waste, too intimately locked into the rock to be extracted.

Gloddfa Ganol Slate Mine

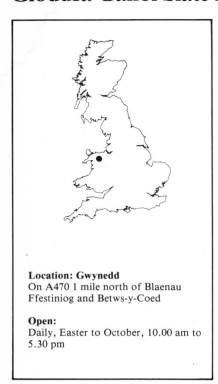

Location: Gwynedd
On A470 1 mile north of Blaenau
Ffestiniog and Betws-y-Coed

Open:
Daily, Easter to October, 10.00 am to
5.30 pm

Rising high above Blaenau Ffestiniog is a hollow mountain from which, for over a century and a half and to all parts of the world, have been sent countless millions of empresses, princesses, small duchesses, wide ladies and narrow ladies. This is not the opening of some Welsh folk tale, but an introduction to the old Oakeley Slate Quarries, the largest in the world, and the names describe different sizes of tile.

Slate has been worked here on thirty different levels, the highest about 1600 feet above sea level, the lowest roughly 500 feet below the level of Blaenau Ffestiniog – a quite staggering range. The huge grey-blue tips of slate waste on the surface bear witness to some great working, but only underground can some idea be had of just how great it is. Thanks to a level tunnel driven through the mountain to simplify the extraction of slate, we can visit the workings of Gloddfa Ganol – the Middle Quarry of the Oakeley concern.

Slate was formed by the wearing down of even more ancient rocks, the silt being deposited on the floor of the ocean which then existed here. The sea bed of sand, silt and mud grew thicker, and its lower layers were subjected to immense pressure and heat. Later earth movements distorted some of these layers, producing a shale which could be split into thin smooth sheets – an ideal material for roofing houses and wherever else a hard-wearing and flat mineral was needed.

When the ever-increasing amount of unwanted rock that had to be removed before the slate could be extracted made the surface quarrying of slate uneconomic at places such as Dinorwic and Nantlle, attention turned to working the slate veins underground by mining techniques. It was then that Oakeley Quarries came into their own, separate workings eventually joining up into a vast maze with 42 miles of tunnels.

Localized collapses were not unknown in the workings, but in 1883 there was an appalling one. Faced with financial problems, one of the three operating companies had been taking the easily obtained slate from the large walls left to support the stone above the mines. At 8 o'clock on the morning of 16 February these walls, perilously thinned down, could support the weight no longer. It is estimated that over 6 millions tons of rock collapsed into the workings and quarry, but, incredibly, no one was killed.

Mass production of roofing tiles and growing wage bills depressed the slate mines. Oakeley was closed in the late 1960s, and, with the pumps silent, the great lower chambers slowly flooded. But slate is once again worked here, by the Ffestiniog Slate Company, though on a much smaller scale than before and by open quarrying.

Families with small children can restrict their excursion to small workings very close to the surface, fancifully festooned with gnomes and the like. But the main tour starts with the long

Above Gloddfa Ganol Slate Mine.

Left Slate on a spoil tip at Gloddfa Ganol Slate Mine.

walk along Hollands Main Tunnel, driven through granite to meet the deep workings in the band of slate known as the New Vein. As the roof is low in places, helmets are issued to visitors at the start. At the end of the long tunnel, which is roofed with corrugated iron in two places where water drips incessantly down old shafts to the surface, a left turn leads into a narrower passage which soon opens up into Chamber 10, the first working.

Chambers in the mine – numbered from east to west – are about 40 feet wide, as is the rock wall left between each to support the roof. Here in 10 is the first sight of the tramway rails, with a gauge of about 2 feet, passing through the next wall under protective timber shuttering.

Beyond the wall the level opens on both sides in Chamber 11. The intrusion of hard bands of rock into the slate here forced the miners to work this chamber in three separate sections, and a continuation can be seen beyond the low wall on the left – Chamber 11 South, its floor some 80 feet below. In the other direction, the roof of Chamber 11 North is illuminated by floodlights.

The route is now straight on into Wall 12 but, once in the wall, it turns sharp left to follow the general lie of the wall through a damp, dripping section. At a junction, one passage contains a slab doorway and the notice 'caban'. Situated at various convenient spots in the whole vast mine complex, the caban originally provided a sheltered place for miners to eat or drink tea in sealed-off sections of old passage or specially built huts. But over the years they took on the status of an institution, each holding formal meetings with minutes taken, and played an important part in the social life of the men.

Past the caban, on the left, is the fenced rim of the chamber seen earlier. Following the left-hand branch of the tramway at a junction, we are on the flat level between this falling-away on one side and the opening which takes you into the third and last part of Chamber 11. Here, down the steeply sloping face which rises in front of you, still hang the metal chains from which the rockmen – the *creigwyr* – would

work, a turn of the chain taken round one thigh. At the foot of the face is one of the flat wagons used to transport large, single slabs of slate.

The next chamber, 12, has been worked very little at this level, but from a viewpoint you can look down a great grey slope to Floor C, about 300 feet below. There is another spectacular spotlit view, too, from Chamber 13. Here the tramway is quite narrow and from it, over a low wall, are seen three giant steps – marking lower working levels – down to the floor 240 feet below.

The next stretch of passage is longer than expected, taking you not only through Walls 14 and 15 but through the section where Chamber 14 should lie. Chamber 14, both North and South sections, does exist, but does not quite reach the level we are following; the small South section can be seen from a tunnel, which may be followed back from Chamber 15.

Here, in 15, the tramway track lead-off to the right was used for dumping waste into the large North section, filling up the old used area below. The open-ended waste-wagon was pushed to the very lip of the hole where a restraining chain suddenly checked it, causing the contents to shoot out. In the roof of Chamber 15 South – which was worked extensively because of the fine grain of the slate – the narrow shaft from the floor above, from which the excavation was started, is floodlit.

The tour finishes in Chamber 16, where you can see a restored tripod crane, once used to lift large slabs on to the wagons for the journey to the surface. The level you have followed continues for a short distance further, but ends in a dangerous collapsed area.

On the surface at Gloddfa Ganol are various exhibits and demonstrations of the other side of the slate story – its cutting and splitting. So intimately bound together were the mining and dressing of slate that men worked as partners, one underground, the other on the surface preparing the tiles from the blocks sent up.

If your walk along the fifth level has given you an appetite to see more of the fascinating world of the hollow mountain, special 'Safari Tours' can be booked at the Craft Shop. These conducted tours are on Floor 11; the old workings there have not been specially prepared for the public, so miners' lights are issued together with helmets.

Llechwedd Slate Caverns

Location: Gwynedd
Off the A470, just north of Blaenau Ffestiniog

Open:
Daily, 26 March to 30 October, 10.00 am to 6.00 pm; last tour of the mine starts at 5.15 pm

Above Leaving the daylight for the depths of the great Llechwedd Slate Caverns.

There are two reactions to the soaring tips of slate waste that rise darkly above the Welsh town of Blaenau Ffestiniog. There are those visitors who see them only as an unpleasant blemish made by man on the natural landscape. And there are those who assimilate them without difficulty: smaller mountains among the great, both dark-clad witnesses to the subterranean workings which threw them up – workings of nature, workings of man.

It is to the latter category of visitor that a trip into the depths of the Llechwedd Slate Caverns will probably most appeal, and on whom it will create the longest-lasting impression. But then again, a visit will provide the more sceptical with a novel insight into the great efforts that have gone into mining and quarrying slate in this hard landscape, and may, just possibly, lead them to a reappraisal.

In the mid-eighteenth century, Blaenau Ffestiniog was simply the name for a group of cliffs several miles to the north of the village of Ffestiniog. Great changes started in a small way, in 1755, when Methusalem Jones discovered slate and started working what was to become the Diffwys quarry, 1 mile to the east of Llechwedd.

In 1833, John Whitehead Greaves set out on foot from his Warwickshire

home for Liverpool, from where he was to emigrate to Canada. Taking one of those quite unforeseen departures from a plan that seems to mark the course of so many pioneers' successes (and failures), he decided instead to take out a lease on two quarries. On both of them he lost money, but he became convinced – having studied the strata of slate at various workings – that a rich vein lay beneath the bleak moorland at Llechwedd. It was a hunch that very nearly broke him financially, and certainly would have if two families of quarrymen, who shared his faith, had not agreed to work for some weeks for no pay. Greaves' gamble, and that of his workers, paid off: one of the final, desperate exploratory tunnels struck the famous Merioneth Old Vein.

Today, Llechwedd is the largest working slate mine in Wales, and in the early 1980s welcomed its two millionth visitor since the miners' tramway opened to the public in 1972. The tramway is one of two separate underground tours available; the newer alternative is a journey into the Deep Mine.

The miners' tramway – which follows the same level throughout – has a 2-foot gauge, like all the Blaenau Ffestiniog industrial tramways and the Ffestiniog Railway. The journey starts behind the ticket office, the track head-

ing straight into the mountain. The four-seater car, one of half a dozen in each train, was specially made for Llechwedd, and is drawn by a 3½-ton battery locomotive.

The first workings passed, on both sides, are small trial excavations, never pursued to any great size. Then comes a brief view of daylight again as the train skirts the edge of the modern open-cast quarry. This has been worked since 1935 to extract the excellent slate left in the giant supporting pillars of the nineteenth-century mines. Back in the darkness, the train now passes two huge chambers on the right. These connect behind the pillar you pass, presenting an awe-inspiring gulf 200 feet deep, well deserving the name given by the old miners: the Cathedral.

A short distance further is the first stop, at a chamber opened up in 1867. Here, Victorian working conditions are recreated by means of life-size dummies, which give a sensible scale to the 75-foot-high opening. This whole chamber was opened up over many years by just two men, helped by one or two apprentices known as rubbishers.

But its production was poor, and only one other member of the team was required on the surface to dress the slate, instead of the usual two: a splitter and a dresser. Records show that only 90 tons of finished roofing slates came from this chamber each year, compared with, for instance, the 240 tons once originating from the next chamber on the tour.

The figure high on the rock face shows the basic method the miners used to secure themselves, by wrapping a tethered chain around one thigh. After the mid-19th century, other explosives replaced gunpowder; but gunpowder is still preferred to this day for most slate quarry work because it shatters the rock less. In the early days, of course, the holes for the gunpowder were bored by hand, three men being capable of drilling about 8 feet each day. Power drills brought a great increase in productivity; now two men can bore more than four times as much in the same period.

Now the train is boarded again, and the journey continues still deeper. Passing a sealed chamber on the right, the train eventually turns right into Choughs' Cavern, which takes its name from the birds that once inhabited the sheer-walled working when one corner was excavated to the open. If the day is bright, shafts of sunlight illuminate the cavern quite magnificently.

The train now reverses out of Choughs' Cavern to the final stopping-point, at the foot of an old incline. Slate blocks, weighing perhaps 2 tons each, once travelled up this to Floor 5 Mill three levels higher, on trolleys raised by means of counter-balanced trolleys filled with water.

Another item of interest here is the caban – the shed. There were cabans in every slate mine, and miners held formal meetings there. The caban formed a centrally important part in the social lives of mine-workers, being run on the lines of an exclusive club, with its own stringently observed rules. As the tea was brewed and drunk, conversation ranged over every topic touching the lives of the miners, from politics to marriage. Following the Welsh tradition, each organized its own Eisteddfod.

Now it is time to board the train again for the return to the surface. But just before you do, spare a moment to place yourself in the proper context of the great mine. You have travelled on only one floor; below you are ten similar levels, and above are another five, occupying a total worked height of 1000 feet. And on each level, chamber lies beneath chamber, pillar beneath pillar, so ensuring the stability of the whole immense working.

The second journey at Llechwedd is a descent into the Deep Mine. A special twenty-four-seat carriage had to be designed for the trip, able to operate within the existing incline, which drops sharply at a gradient of 1 in 1.8. The 3-ton carriage is hauled by cable and incorporates two fail-safe braking systems, one cutting in if the speed of descent exceeds 2.4mph – hardly pell-mell. The incline carries you to the dis-embarkation-point two floors below the surface. The concrete lining passed on the first part of the descent seals off the Old Level (Floor 1).

The tour takes you through a looping network of passages and chambers, one of the highlights of which is the dark waters of the 50-foot-deep lake into which the slate waste tumbles. Unlike the tramway tour, which concentrates on mining techniques, the Deep Mine tour has been planned to give an insight into the social life of the miner, and the growth over the years of Blaenau Ffestiniog. There are ten narration-points at which our unseen guide – Sion Dolgarregddu, a young apprentice who started his working life at Llechwedd in 1856 – tells us his reminiscences via the voice of an actor from Theatr Clwyd. Much use is made of both light and sound in the various presentations, as well as music from the Welsh harp and songs from the Moelwyn Male Voice Choir.

There are a number of surface exhibits at Llechwedd which add to an understanding of the mining story. The ride on the miners' tramway ends in the large Old Mill, in which the power transmission shaft – once driven by a huge overshot waterwheel – runs the full length. In the Slate Heritage building an audio-visual programme is screened every half hour, Emlyn Williams narrating the story of Welsh slate from Roman times to the present.

Right Splitting slate in Llechwedd Mine.

Dinorwig Power Station

Location: Gwynedd
Off the A4086, 1 mile east of
Llanberis

Open:
By advance postal booking only, to:
Station Manager, Dinorwig Power
Station, Llanberis, Caernarfon,
Gwynedd; telephone 0286–870166.
Unbooked visits may be available
from 1985

The coalfields of South Wales have provided Britain with much of its power for many years, but the thought of power coming from North Wales – the land of bleak mountains and the occasional gash of some expired slate quarry – would have been laughed at only a handful of years ago.

Now, buried deep in the rock at the foot of the Llanberis Pass, from which countless thousands of walkers and climbers have set off for the ascent of Snowdon, is a great power station, the innermost workings of which can be viewed by the public. This power station – constructed beneath the old Dinorwig slate quarry, worked for 160 years – is unique in being the largest pumped storage station in Europe, and the first in the world designed to deliver from zero load to full load in only 10 seconds.

Unlike conventional hydro-electric power stations, a pumped storage system needs both an upper and a lower reservoir. During the day, when demand for electricity is at its highest, Dinorwig takes water from its upper reservoir – Marchlyn Mawr, a natural lake – at up to 85,000 gallons every second. Having passed through the turbine generators, the water is discharged into the lower reservoir, another natural lake, Llyn Peris. At night, the generators revert to their second function as pumps, and restock the upper reservoir with water from the lower.

Put as simply as this, it does sound like a case of robbing Peter to pay Paul. In fact there is a substantial benefit for the Central Electricity Generating Board, for Dinorwig acts as a huge 'buffer' for the national grid. During the night the electricity needed to pump the water back up is drawn chiefly from coal-fired power stations that would otherwise be running uneconomically at a low load. They are kept running at a high load, and the surplus electricity produced at night by these stations is used at Dinorwig to pump water back up the mountain. During the day, Dinorwig can return this stored power to the national grid, and does so most effectively – because of its ability to respond rapidly without any start-up and stand-by costs – by supplying power to cope with sudden peaks in the demand or with the failure of supply from some other station.

To create a power station underground capable of dealing with this enormous quantity of water has involved an engineering project of quite spectacular size. The tunnel complex around the huge chambers of the generating and transformer halls puts Spaghetti Junction to shame. A gently sloping low-pressure tunnel leads from the upper reservoir to a vertical high-pressure shaft just over 1 mile distant. This shaft, over 30 feet in diameter, drops sheer for over 1400 feet to the high-pressure tunnel which takes the water the final 2300 feet to the generating hall. This tunnel feeds six smaller

Above The view across Llyn Peris to Elidir Fawr.

Right Main inlet valve gallery of Dinorwig Power Station.

tunnels, each leading to a pump-turbine, and each of these has its own main inlet valve – or tap. Some tap. It works like a conventional ball valve, but in this case the valve is controlled by two arms bearing closing weights of 16 tonnes each. These gigantic controls, which work on a fail-safe basis, can shut off the great rush of water in 20 seconds and open in a mere 5.

The pump-turbine and generator-motor assemblies are, of course, on an equally impressive scale – the weight of each generator-motor's rotor (the spinning part) alone is 445 tonnes. Compressed air is used to start these rotating when there is a possibility of power being needed at a few seconds' notice.

How much of this will you see on your tour? This depends on operational requirements at the time, but most visitors will see the main machine hall, 588 feet long, 77 feet wide and 195 feet high; the main inlet valve gallery, 480 feet long, 21 feet wide and 60 feet high; the draft tube valve gallery, where the six 12-foot diameter valves control the tunnels leading to the lower reservoir; and the pump-turbine floor through which the massive drive shafts lead up to the machine hall above. Whenever possible, the central control room can also be viewed. This automated nerve centre of the whole operation, with its sprinkling of control consoles and display boards, makes a sharp contrast with the awesome scale of those giant machines it controls.

NORTHERN ENGLAND

The great chain of the Pennines, often referred to as the backbone of England, accommodates two of Britain's main cave regions: the Yorkshire Dales and the Derbyshire Peak District. Of these, the Dales stand paramount, offering the caver the greatest number and variety of caves in the whole country. Whereas Yorkshire's caves could be described in a small pocket book in the 1960s, such has been the rate of discovery and exploration that six full-size volumes are now required.

This is the land of potholes, their great shafts plunging sheer into the limestone, many still very active and so offering the added thrill of vigorous streamways and waterfalls. Although some caves can achieve considerable depths by descending steadily over a great distance, the feeling of depth in a Yorkshire pothole, when shaft after shaft has to be descended, can be most profound.

In view of the richness of Yorkshire, it is perhaps surprising that only three show caves operate regularly in the region (though others have come and gone in the past): Ingleborough Cave, Stump Cross Caverns and White Scar Caves. However, these few are first-rate, and give the visitor a true taste of the special appeal of these great Northern cave systems. They are also situated in some of the grandest limestone country in Britain.

At the southern end of the Pennines, in the Peak District, is the greatest concentration of show caves and mines in the entire country. Concerning the ten sites in the county of Derbyshire, the most telling statistic is that no fewer than eight have involved mining at some point: but these are by no means just mines; in more than half, miners, while excavating lead, fluorspar or Blue John, also discovered natural caves.

The variety of sites here is considerable. You can take an underground boat trip along a low tunnel to enjoy the spectacle of the Bottomless Pit in Speedwell Cavern, or walk into the gaping maw of Peak Cavern, whose 100-foot wide entrance once sheltered a busy rope-making community. In Blue John Cavern and Treak Cliff Cavern you can see where the beautiful Blue John mineral has been mined from the only deposits to be found anywhere in the world, or you can take a cable-car across the Derwent Valley to the Heights of Abraham where two caves-cum-mines can be inspected.

Poole's Cavern is a natural cave that the public has toured for centuries — certainly as far back as 1582, when Mary Queen of Scots bestowed her patronage. Goodluck Lead Mine at Via Gellia provides an underground tour of an old lead mine as well as a number of fascinating surface exhibits.

Two other northern sites are quite different. In Staffordshire, the Chatterley Whitfield Mining Museum gives the visitor the opportunity to drop in on a modern coal mine 700 feet underground. To the east, in Nottingham, networks of intriguing man-made passages are cut into the readily-yielding sandstone. Those under the castle are open to the public, as well as tunnellings in several other parts of the city.

Blue John Cavern and Mine

Location: Derbyshire
1½ miles west of Castleton on the A625. If the Mam Tor road is closed at Castleton, drivers travelling west should go via Winnats Pass

Open:
Daily, Easter to 31 October, 9.30 am to 6.00 pm
1 November to Easter, 10.00 am to dusk
Guided tours, which last about 45 minutes, start every 15 minutes

This cave and mine pierces the only hill in the world in which Blue John stone is found. Such a rare and beautiful mineral merits a short description for those unacquainted with its undulating bands of blue, purple, white and yellow. Differing from other fluorspars in its distinctive colouring, Blue John was probably formed when hot gaseous material was forced up from the earth's core into the cracks and joints of the surface crust where it combined chemically with the surfaces of the limestone. There has been much speculation about the cause of the bands of gorgeous colour, and suggestions have ranged from the intrusion of asphalt, manganese dioxide or calcium permanganate to rare earths.

Whatever the cause, Blue John has been highly prized for centuries. The first known mining was by the Romans, 2000 years ago, and vases worked from the mineral — two of which were unearthed in the ruins of Pompeii — brought huge prices. Of the fourteen known veins of Blue John, eight have been worked in this mine for centuries,

though for the period from the Roman occupation until 1700 no records have been found. The popularity of the stone was re-established in the late eighteenth century, and considerable numbers of people were employed to mine and turn it.

So to the mine, and the natural caverns discovered by miners nearly 300 years ago. The tour starts with a short descending flight of steps, through a mined passage for some 20 feet, from the foot of which you go at once into the beginning of the natural caverns with their formations and mineral coloration. At this first horizontal level, the original rock joint responsible for the passage formation can be followed. Here also are the first of the mine workings: Roman Level and Five Vein Working. A natural pothole led to the lower levels, and the miners took advantage of this to construct a stair-

Below Amidst the green of the Peak District, the entrance mine-cum-cave of Blue John.

Blue John Cavern and Mine

way down it to the fresh veins of Blue John below. Notice the distinctive coloured light halfway down – you will see it again looking *up* the pothole, providing some perspective.

From the foot of the pothole, you follow the course of the river, long since vanished. Much of the floor levelling here was done by the miners disposing of small spoil (refuse) from the workings. Fortunately, though, they had to make a number of spoil heaps in various places, for it is from these that much of the Blue John is recovered today for the small pieces used in jewellery. The cost of starting to mine Blue John again from veins on a large scale would be prohibitive.

As you follow this passage, keep an eye open for the inverted potholes in the roof, formed when the entire passage was flooded by water under some considerable pressure. Now comes the Bull Beef Working from which some of the largest and best pieces of Blue John have come. So densely coloured were the Bull Beef minerals that the pieces had to be heated very carefully to reduce the colouring – but too much heat and the painstakingly mined Blue John was reduced to a worthless white. A few yards past this working is the start of the caverns proper, and the lofty dome-shaped Grand Crystallized Cavern with its pronounced colouring and crystallizations. The tallow-candle chandelier and windlass here were used to illustrate the chamber for tourists in the days before the cave was electrically lit or, earlier, illuminated by acetylene lamps and paraffin wax candles.

The Waterfall Cavern follows immediately, the whole of the left-hand wall covered in rippling flowstone – an ages-old 'waterfall' frozen in a trice. Rising from one wall shelf is the stumpy Beehive Stalagmite, and, contrasting with the brick-red ceiling from which it hangs, dangles the long and slender Sword of Damocles stalactite. In the next chamber, Stalactite Cavern, are a distinctive stalagmite and, from the roof, here resembling a meandering river bed turned through 180 degrees, a fringe of stalactites.

The passage continues on a now rather winding course to the large Lord Mulgrave's Dining-Room, named after a meal he is said to have put on there for local miners. Two watercourses met here, almost at right angles, producing the pronounced whirlpool cutting effect on the walls. The uppermost

Above The Mirror Lake in Blue John Cavern.

Right The Dining Room in Blue John Cavern.

effect of this action is seen in the clearly marked horizontal line round the walls. In a corner of this chamber is another working, the old Dining-Room level, and here you can see Blue John in its natural state, locked into the limestone, showing both a vein and a cross-sectioned nodule of the material. The small opening in this working leads to a much larger cavity beyond in which the miners dumped many hundreds of tons of spoil.

A short distance further is the Variegated Cavern, the last of the main series of chambers shown to the public. Rising to a height of over 200 feet, it is a fitting climax to the tour. It derives its name, obviously, from the variegated patches of minerals coating the walls: reddish-brown iron oxide and the much darker manganese dioxide. The whole left-hand wall is coated with stalagmitic formations, and in a cavity opposite the terminal platforms is a fine translucent creamy curtain.

On the return journey to the surface, in the section of the cavern that is completely dry, the rock glistens with a myriad crystals – and then the sudden surprise of the Balancing Rock. Estimated to weigh 20 tons, it is com-pletely detached from the walls, only touching at two points. Some take its precarious balancing act to indicate the lack of local earthquakes or tremors over the past 80,000 years or so.

The main caverns rejoin the exit passage near the entrance to the Organ Room working, famed for its richly coloured Blue John, but never yielding a vein much more than 2½ inches thick. At the bottom of the pothole again is a door leading to the Dining-Room working from which was recovered yet another distinctive form of the mineral, much valued in the making of chalices and vases.

On reaching the surface, if you have time to spare, visit the Cavendish House Museum in Castleton, with its Ollerenshaw Collection of exquisitely-worked Blue John.

Treak Cliff Cavern

Location: Derbyshire
1 mile west of Castleton on the A625.
If the Mam Tor road is closed at
Castleton, drivers travelling west
should go via Winnats Pass

Open:
Daily, Easter to October, 9.30 am to
6.00 pm
Winter months 9.30 am to 4.00 pm
(except Christmas Day)

The Peak District of Derbyshire occupies some 540 square miles, yet — through a quirk of geology and the particular attention of miners in the past — four of its most famous show caves lie within a 1000-yard radius, just west of Castleton. Although the 'baby' of the quartet, in terms of the passages so far discovered, Treak Cliff Cavern treats the visitor to what cavers themselves regard as some of the finest stalactites in Derbyshire.

Walking up the zig-zagging path leading from the car park to the entrance, from which there are excellent views of the Hope Valley and the surrounding hills, we are treading in the footsteps of the miners who worked part of the cavern for Blue John stone 200 years ago. The hill which Treak Cliff Cavern pierces stands unique as the only place in the world where workable veins of Blue John are found — a rare form of the mineral fluorspar. Banded with layers of colour generally of a purple or blue basis, but sometimes white or even yellow, Blue John has long been valued in jewellery and ornaments. Just how the colouring was formed still vexes the scientists, though there are various theories.

When the cavern was first mined for Blue John is not known precisely, but certainly by 1762 miners were working in the first of the two series, or parts, of the cavern — the Old Series.

The Old Series rises steadily throughout its length of about 600 feet, steps being installed in the steeper places. Its passages intertwine at various levels, but the path takes a more or less straight route through them. The water which formed them is long gone, now working at a deeper level.

In the chamber about 100 feet in can be seen a vein of Blue John. A great deal of top-quality Blue John was mined here in the First World War, going the way of ordinary unmarked fluorspar as a flux in blast-furnaces — an ignominous end for such a beautiful mineral. In addition to the mining in the Old Series underground, Blue John was taken from shallow quarries on the surface. It was from one of these, in 1923, that miners broke into a very small cave containing several human skeletons and flint implements dating probably to the Bronze Age.

Back in the cavern, further up the trail is the Fossil Cave. Here the water has etched away the limestone but left large quantities of fossils standing out in relief. Those in the roof which look like screws (the old miners called them screwstones) are the remains of crinoids, a life-form not unlike the starfish, extant when these limestone beds were first forming on the ocean floor a third of a billion years ago.

The highest point of the tourist route comes in the Witches Cave, the largest chamber in the Old Series, 80 feet long and with a width and height of about 20 feet. Much of this chamber was filled with clay deposits until 1950. When these were cleared to give a better impression of the chamber's size and shape, previously unknown veins of Blue John were revealed in the process.

For the second, and most breathtaking, part of the tour, we must thank an accidental discovery by the miners in 1926. They were blasting in the part of the Old Series that continues rising from the Witches Cave, and in a spot quite close to where the Bronze Age skeletons were found on the surface one charge opened a hole in the floor, dropping steeply down into a chamber. Using ropes, four miners clambered down and explored the remarkable chambers which you are about to see. Unfortunately for them, but fortunately for today's visitors, no new veins of Blue John were found, so there was no need for the blasting and mining that would undoubtedly have spoiled the treasures of what is now called the New Series.

And it was another accident which revealed a straight-forward route into the new section, avoiding the steep drop negotiated by the original explorers. In 1932, six years after mining had ceased, some miners who had worked the Old Series returned to recover some Blue John which they just happened to remember, had been buried under a clay fall. Their excavations opened up the passage which you will now follow — a discovery which led to the cavern being opened as a show cave in 1935.

From the Witches Cave we follow steps down and through a doorway to the point at which the full beauty of Aladdin's Cave is spread before us. Eighty feet long and 30 wide, the chamber rises to a height of 50 feet, the ceiling decked with stalactites. The black hole visible as you descend through the boulders on the floor is the entrance down which the miners first climbed. Some of the boulders are capped with stalagmite formations. One of the largest groups is the Seven Dwarfs, the tallest of whom has taken about 70,000 years to reach his present stature of 18 inches.

On the same level, and just around the corner, is Fairyland, with its clusters of exquisite formations on the wall.

The high passage veers to the right, taking us into the breath-taking Dream Cave, a chamber festooned with hundreds of stalactites. The longest, hanging nearly 4 feet, comes to within 1½ inches of the stalagmite below. The whole configuration is aptly named The Stork, but it will be another 1000 years before this bird finds firm footing when the two formations finally join.

Immediately ahead, the 40-foot-high Dome of St Paul's provides the most colourful finish possible, the formations coating its walls being tinted by overlying minerals into a rich spread of the spectrum. In all likelihood the cavern continues beyond this point, but a huge block of limestone plugs any way on and explosives cannot be used for fear of damaging the rich formations which make this such a memorable cave. So the return to the surface begins, the actual exit being made from the Old Series through a passage almost level with the Witches Cave.

Right The Stork Stalactite, Treak Cliff Cavern.

Speedwell Cavern

Location: Derbyshire
½ mile west of Castleton at the foot of Winnats Pass

Open:
Daily (except Christmas and Boxing Days), 9.30 am to 5.30 pm; tours start every 15 minutes

Although not possessing the rich stalactite and stalagmite formations of other show caves, Speedwell Cavern, in the heart of the Peak District, has the distinction of offering a tour the major part of which is by boat. Begun as a lead mine in 1771, Speedwell Cavern was abandoned by the miners some twenty years later because of the small amount of ore recovered — poor returns for their enormous labours in driving the underground canal. Fortunately for us, the canal remains, giving us access into the depths of the hillside.

Once through the entrance door set in a high limestone wall, visitors must use their legs for at least the first part of the tour, for here — doubling back underneath the road — is a flight of 104 steps. At the bottom comes the easy part, the embarkation on a boat which the guide propels by pushing on the rock walls with his hands. No unnecessary mechanization here! The tunnel is 7 feet in diameter, and the canal 3 feet deep.

The first feature of interest along the canal is the Poor, or Little Winster, Vein, with lead ore and yellow fluorspar crystals, hardly 1 inch wide. So poor did it prove, in fact, that the miners worked it for only 120 feet before abandoning it. Close by are the small pockets in the rock, 1 foot or so wide, which rewarded the miners with blocks of lead ore embedded in the clay. Some 300 feet beyond the Poor Vein and 450 feet below the surface, a short side canal goes off at Halfway House to the right to the Longcliff Vein workings. These are now walled up but are said to extend for ½ mile on that side. Further, and two more veins are passed, but these are small also and contain mainly spar — once used in optical instruments because of its peculiar optical qualities. If a ½-inch-thick section of spar were placed over this page, each line of type would be seen as two.

The remaining half sections of the miners' drill holes used for blasting can be seen for much of the way along the canal. Each was about 18 inches long and 1½ inches in diameter, and would have taken a pair of miners (one holding the drill, one hammering) at least a couple of hours to make, two pairs of miners working side by side. When up to twenty had been completed, the holes were filled with gunpowder, plugged with clay and fired by straw fuses. If it was impracticable to retreat to a side working, the miners sheltered in a shallow 'safety' hole in the tunnel wall during blasting. It seems that as this part of the tunnel was driven to its conclusion ventilation became a problem, so a bellows hole was excavated near the Halfway House. Here a boy worked the ventilation bellows on an eight-hour shift.

Having travelled some 1500 feet in the boat, through the close confines of the tunnel, the contrast as you enter the Bottomless Pit Cavern is the strongest possible. This is where the miners struck a natural cavern, the roof of which can be seen only by spotlight as it curves up out of sight 140 feet above. Having stepped from the boat on to the stone-built platform, one is drawn to the hole opening up on the other side of a safety railing. This is the Bottomless Pit, the floor of which the spotlight now picks out, some 70 feet below. This hole is claimed to have swallowed 40,000 tons of mining rubble without its water level being affected.

Right Bung Hole, Speedwell Cavern.

Left The mined passage beyond the Speedwell Show Cavern.

Speedwell Cavern

Above The illuminated Canal in Speedwell along which the visitors are ferried by boat.

Left The Bottomless Pit, marking the end point of the tourist excursion.

As you stand here, reflect on one of the quirks of caves in terms of depth measurement. You are now some 840 feet below the surface of the hillside, yet have descended only that one flight of steps. Because of the anomalies that can arise if depth measurements are taken from the surface directly above a certain point in a cave, cavers note depth readings as the distance actually descended from the entrance.

The journey into the depths of the

Peak District ends here, but first the guide will point out the low arched entrance to the Far Canal – the continuation of the mined level. There is the distant rumble of a waterfall. The water level in both the Far Canal and the one you will again traverse on your return is maintained by the rock platform and a dam in the Far Canal. Examination of the few remaining documents relating to Speedwell Cavern suggest two interesting probabilities: that the miners had a top entrance (now lost) to the Bottomless Pit Cavern and thus knew exactly where to drive the tunnel; and that the deliberate flooding of the tunnel was an attempt to mine lead as in the Duke of Bridgewater's coal mines at Worsley, where a canal led right up to the coal face.

The total explored length of the mines, and the natural caverns beyond the Far Canal, is about 2½ miles. Cavers use inflatable dinghies to gain access to the inner reaches, wading where the water level permits. At one point, the Whirlpool, an ebbing and flowing stream, enters the passageway, raising the water level quite noticeably, and without warning, alarming any cavers not aware of the situation. The author well remembers a trip past this point, each caver nervously eyeing the water level, convinced that the cave was flooding. After several minutes someone voiced their fears, was hurriedly agreed with, and a rapid exit started. Only then did someone – the only caver who had been down the cave before – remember the strange behaviour of the Whirlpool, and the trip was resumed.

139

Peak Cavern

Location: Derbyshire
On the western side of Castleton on
the A625

Open:
Daily, Easter to mid-September,
10.00 am to 5.00 pm

The approach to Peak Cavern must surely rank as the most impressive of any show cave in Britain. From the bridge by the car park the walk leads by the riverside past old lead miners' cottages, often set at angles to one another. The path runs ever deeper into a huge limestone gorge, with cliffs 250 feet high, crowned by the ruins of the Norman Peveril Castle. Turning a corner you are suddenly confronted by the gaping maw of the cave entrance, the Vestibule – 100 feet wide, 60 feet high and 330 feet long.

The floor within the entrance drops away to the left in six large steps. Each of these was a rope-walk where, for several centuries, rope was spun. The last rope-maker retired in 1974 at the age of eighty-nine, bringing to an end the custom whereby every newly-married couple in Castleton was presented with a Peak Cavern clothes-line. Each rope-walk was a family business, and the last rope-maker recalled that thirty people worked there in his youth. To make the rope the rope-maker slowly walked backwards adding strands of Italian hemp from the hank which he carried on his belt. On the lower walks you can see reconstructions of the string-winding drums and, next to the admission office, an iron device known as a runner, used to make rope or line.

According to an account published in 1700, which referred to the cave by its older and blacker name of the Devil's Arse, many of the poor of Castleton lived in habitations erected in the huge entrance itself. The entrance has never been excavated, and so nothing is known of early man's activities there, although it certainly seems to have been used. A ballad by Ben Jonson written in about 1621 mentioned that medieval beggars held a banquet there each year, led by the robber-king Cock Laurel.

Right Surprise View, Speedwell Cavern.

Below The remains of the rope-making apparatus in the huge entrance to Peak Cavern.

Peak Cavern

Apparently he did not share, or worry about, the many grim legends associating the cave with Hell, and even invited the devil to join in the celebrations.

The tour of the cave starts here, by the large stalactites hanging down an aven in the roof. At the inner end of the Vestibule, the path descends to a gate, then to a small chamber called the Bell House. Here the cold cave air – a steady 47 degrees Fahrenheit throughout the year – is met by the visitor for the first time. In very heavy rain, when the cave is closed to visitors, the water level rises in the cave, sometimes reaching halfway up the gate.

After stooping through the appropriately-named Lumbago Walk, you come to the Inner Styx, a small pool. In the old days visitors were ferried over this into the Great Cave, lying flat-out in the bottom of a boat. Now, a more convenient, if less exciting, artificial tunnel takes you comfortably past this obstacle. The cave contrasts impress-

Left A caver wades cautiously through waist-deep water in one of the inner passages of Peak Cavern.

Below Five Arches, Peak Cavern.

ively with the previous low section, measuring 150 feet wide, 90 feet long and 60 feet high.

A high passage leads to Roger Rain's House with its cascade of water. A small spring that disappears on the surface behind Peveril Castle reappears here. A switch is thrown, and powerful floodlights pick out the Orchestra Chamber, high on the wall. In the past, wealthy visitors were treated to singing from the balcony by the Castleton choir, plus, if the guide felt in the spirit, a small gunpowder explosion. Today's visitors have the quieter but just as strange experience of a few moments with the lights extinguished, and only the unimaginable darkness of the cave. Put your hand in front of your face – you will see nothing but slight overall greyness, and even this would vanish if your eyes were allowed to adapt to the dark.

From the broad Pluto's Dining-Room chamber you descend a flight of steps to the Devil's Cellar and follow this passage until, at Halfway House, you meet another feature named with the devil in mind: the river Styx. This flows into the passage on the left for 120 feet before sumping and continuing its journey to the resurgence near the main

entrance. This is the active part of the cave, and in rare times of high flood soil is carried in and deposited on the walls. In the immediate vicinity of the flood-lights, where there is sufficient heat and light, seeds carried down with the soil can germinate, resulting in the small green patches behind some of the lights.

As you continue along the river passage, four small bridges cross the water at various points, and above the fourth are two very large avens more than 100 feet high. The last part of the passage, Five Arches, ends in an aven between the final two arches which is a perfect half-sphere and, because of its bell-like appearance, takes the name Great Tom of Lincoln.

A few yards more and, at a T-junction, ½ mile from the entrance, the tour is at an end, and you return with the guide along the same route. For the caver, however, Peak Cavern continues for another 2 miles: to the left, following Buxton Water, for only a short distance before a long sump is met; to the right, to a large series of demanding and exciting passages and chambers. In 1959 these were the scene of a tragedy when a caver became inextricably trapped while exploring a narrow gut-like shaft in the floor.

Bagshawe Cavern

Location: Derbyshire
On the southern edge of Bradwell, on the B6049 7 miles south of Chapel-en-le-Frith; the cave is signposted in the village

Open:
Weekdays, 1 April to 31 October, 2.00 pm to 6.00 pm; weekends and Bank Holidays, 10.00 am to 6.00 pm 1 November to 31 March, by advance booking only, telephone Hope Valley 20540/21298

Like a number of other natural caverns in the Peak District, Bagshawe Cavern was discovered accidently by t'owd man – the name cavers give to the lead miners of yesteryear. The discovery was made in 1806 by the miners working for the Bagshawe family of Ford Hall in Chapel-en-le-Frith. Not surprisingly, the cave was named after the family, and the story has it that one of the female Bagshawes went down the cave shortly after its discovery and gave the various features the names that are still used today. The link persists, a Bagshawe descendant still visiting the show cave occasionally.

The old lead mine, which ceased working at some unrecorded time in the early nineteenth century, was known as the Mulespinner Mine, and it is down this – via a stairway of 102 steps – that you reach the natural caves below. From the chamber at the foot of the steps the tourist route extends for almost ½ mile, with a number of decorative features. The Elephant's Throat is an imaginatively named natural chimney with a lining of flowstone,

followed by a flowstone wall highlighted by curtains: the Chandler's Shop.

A narrow rift leads to one of the main features of the show cave, the Grotto of Paradise, a chamber extensively decorated by tinted flowstone. The final section takes the visitor along a completely natural, wide, boulder-strewn stream passage; for once you can get just a taste of a natural cave passage, without the usual concreted and carefully levelled path. Now it is dry, but the waters of winter can percolate in sufficient quantity the overlying limestone to restore the passage to the active state.

So to the end of the show cave, marked by a big, clear floodlit pool with a particularly large stalagmite, and a decorated chimney bearing the name Niagara Falls.

As with many show caves, Bagshawe continues well beyond the routes opened up and prepared for visitors. In fact, its total length is 10,000 feet. On the way in you will have been shown the Dungeon, a large 20-foot-deep pothole in the floor. Using ladders or a rope cavers can descend this to the lower series, which they can follow, through pools, crawls and small chambers, for ⅓ mile eastwards.

The upper series, which continues where the show cave section leaves off, can be visited on special 'adventure caving' trips (by appointment – allow ten days in summer). These give the non-caver an opportunity to experience the very special pleasures of passing through completely natural cave for an additional ½ mile of passage.

Left Flat-out progress for two cavers.

Right Wading through part of the wild cave, the lower part of which is reached via the Dungeon.

Goodluck Lead Mine

Location: Derbyshire
Via Gellia, about 2 miles west of
Cromford on the A5012; park in the
lay-by

Open:
11.00 am to 6.00 pm on the first
Sunday of every month; at other
times (except Saturdays) by
appointment, telephone 0246–72375

Carefully restored over a number of
years, and quite devoid of the sometimes harsher aspects of commercialization, Goodluck is a fascinating
example of a mid-nineteenth century
Derbyshire lead mine. It is well worth
visiting if its (at present) limited
opening hours fit your itinerary.
Helmets and miners' lamps are provided.

The way the mine came into its
present ownership provides an
intriguing story of how ancient laws and
customs still hold today. From far back
miners have enjoyed certain privileges
in their quest for galena – the common-
est lead-bearing ore – in the Peak
District. At first handed down by word
of mouth, they were set on paper in
1288 and continued, with occasional
amendments, until a major revision and
simplification in the middle of the last
century. Administration of these laws is
undertaken by Bar Moot Courts, separate ones being held in the High and the
Low Peaks. A miner wishing to register
a claim on a new vein had to present a
dish of ore (a volume of 15 pints) to the
Bar Master, who then awarded the
miner 2 founder meers. (A meer is a
measure of between 29 and 32 yards.)
These having been worked, the ore
from the next – the Lord's Meer – went
into the pocket of the mineral lessee.

Beyond that all the meers were the
miner's own Taker Meers, except for
every thirteenth dish of ore, which had
to go to the owner of the mineral duties.
In some areas the Church imposed tithe
duties, which were particularly resented
by the miners.

Once he had established a claim, the
onus was on the miner to work the mine
regularly. If he did not, unless he had
problems with bad air or flooding,
another miner could lay claim to the
mine by, quite literally, nicking it. The
owner was given twenty-one days to put
it on a working footing. A Moot official
would visit the mine every week for
three weeks, marking each call by nicking the woodwork at the entrance.
After three nicks had been made, the
new claimant could move in to produce
the first dish of ore to free the mine.

It was exactly by this old custom of
nicking that the present owner, Ron
Amner (a miner by profession, and

fascinated by the workings of his predecessors), acquired and re-opened
Goodluck Mine in the early 1970s.

The walk up from the main road to
the mine entrance (over a stile and a
bridge to where the miners' path
branches off left to the prominent spoil
heap) takes you past a number of relics
of the mine's past scattered over the
level area. Opposite the ore-dressers'
coe (the name for a small building) are
two shafts, the larger fitted with the
simple hand-winch known as a stowes.
With this the ore was raised from the
mine in a bucket, two buckets being
used in a counterbalanced system for
the deeper shafts. The smaller shaft is a
climbing way, used by the miners to
reach the workings; they climbed using
the small holds cut into the shaft's stone
lining – worth pondering on when one
next grumbles about queuing for a bus
or train for work!

The circular building is the gun-

Above The jigging machine which separated the heavy, lead-bearing ore from the lighter waste.

Left The use of a 100ft incline, down which the tubs of ore are winched, overcoming the problems of a rock fault.

powder store, its doorway aimed at the valley rather than the mine workings. Its roof would have been of turf – far less lethal than stone or slate in the event of an explosion.

The wooden device to the right of the powder house is an ore-jigging machine. Crushed ore was loaded into the tray suspended on long handles, then jerked through the water in the trough. The heavy lead ore settled at the bottom of the sieve and the lighter waste material was periodically scraped off. Halfway to the entrance from here is a stone-lined, stepped buddling trough, used to recover the very fine ore dust which accumulated in the bottom of the jigger. The sludge was raked against a small flow of water, the lighter waste being washed to the trough bottom from where it was dumped on the tip. The lead still locked into this waste is the reason why the side of the top is barren of plant life, except for a meagre covering of tough moss.

The mine was started from its present entrance in 1830 to intersect the very rich Goodluck vein, and yielded ore until the 1860s when only the scrins (small veins) parallel with the main vein were left to exploit. The mine fell derelict at the turn of the century, was briefly revived during the First World War, then worked intermittently – for barytes as well as lead – until 1952. In that year the entrance collapsed, remaining sealed until the early 1970s when the owner, having staked his claim, rebuilt it.

The first few yards of the entrance passage through heavily fissured rock were very easy to drive. Beyond that, blasting was necessary. Examples of later drilling mechanisms are displayed in the first section, but those who drove this particular mine had to drill the shot holes by hand, each one taking about two hours. Two or three ounces of gunpowder were forced into the hole, then stemming – usually clay – was rammed home to fill the rest. A hole was pierced through the stemming with a long metal taper, and a crude, gunpowder-filled straw fuse inserted.

After 270 feet the passage intersects the Silver Eye Vein, then, after another 60 feet, the Black Rake Vein. Both these were being mined from the surface when this passage was driven, and it says much for the degree of co-operation between the three concerns that both earlier mines were allowed to take out their tubs of ore through the new level (for a consideration). The chutes

opening out into the passage at these points were used to drop ore into tubs.

A short way after the Black Rake workings is an incline, rising 30 feet over its 100 feet-length, built to surmount a fault. In other mines in the area a transfer shaft was excavated at this fault, but the incline is much more efficient as the ore could simply be winched down in a single tub rather than having to be transferred from one to another.

Shortly after the winch at the top of the incline, the passage suddenly intersects the Goodluck Vein and the mine begins. You can see here the wide vein of creamy barytes flecked with galena and, to the side, a much smaller parallel vein – a scrin. We follow the vein to the right for a stretch of 100 feet, which was worked by the earlier Moor Jepson mine, to signatures of miners scribed on the wall – presumably to celebrate the completion of the crosscut in 1831. From this point, Warl Gate passage heads off to the south-west, intersecting four parallel veins: First Warl Vein, Middle Warl Vein, Upper Warl Vein and Holmes Vein. These were worked in the 1840s.

The shaft in the floor opposite the carved initials was made to test the depths of both the Warl Gate and Goodluck veins, and then for working the latter. Filled in during the 1930s, it is being excavated to reveal the lower workings in Goodluck once more.

After another 100 feet or so is the entrance to Gulph Gate, a level driven off for 400 feet to meet Godbers Scrin, encountering on the way the three Warl veins again. On the floor of Gulph Gate is a section of wooden tramway with a reconstructed wooden tub and, opposite, a display of the various mining implements found in these workings.

Further along is a shaft in the ceiling of the passage. This, known as a raise, was part of the original access route into the Dales Founder Mine, the miners climbing up the wooden stemples wedged across each side of the shaft. A short distance further – past a display of geological specimens – a chain ladder gives a different means of climbing another raise, whilst a third raise has hanging down it a climbing chain, probably the most strenuous method of all.

At a distance of 1220 feet from the entrance comes the Goodluck Forefield, the point where the vein finished after being worked for 700 feet. Various crosscuts were driven from the main working, but none gave the same level of yield as the rich Goodluck Vein.

Royal Cave

Location: Derbyshire
Temple Road, Matlock Bath; 100 yards off the A6, from which it is signposted

Open:
Daily, 1 April to 31 October, 11.00 am to 5.00 pm
1 November to 31 March (except Christmas Day), one tour daily at 2.00 pm

Royal Cave is part of a much larger complex of old workings and natural caverns with such delightful names as Hopping Mine and Tear Breeches Mine, sources of fluorspar and calcite. In the mid-nineteenth century, when Matlock Bath enjoyed a wide reputation as a spa town, the many people who came to take the waters were treated to a choice of no less than ten show caves in the town, one being New Speedwell Cavern in this mine complex.

Fluorspar was last mined here as late as the 1950s, when open-cast quarrying was used; hence the quarry in which the entrance building is sited. Royal Cave (to use its present and presumably permanent name: it has been re-christened a number of times) was largely forgotten until the 1970s, when a careful survey showed that it could be redeveloped successfully as a modern show cave. It was opened in 1980.

Unfortunately mining has removed most of the fine natural decorations, such as the fluorspar waterfall that had pride of place in the old Speedwell show cave. As a result, it was decided to concentrate on developing a show cave with a series of elaborate, computer-controlled *son et lumière* displays.

Before the tour proper, visitors are given an audio-visual presentation in the entrance building, the interior of which is used to extend the underground theme. The slides and commentary deal with the development of mines over the centuries, the uses to which man has put caves, and the history of Royal Cave itself. In the cave, seven scenes depict a variety of subterranean activities at different periods of history. One tableau shows early man painting the walls of his cave, another the scene of a mining disaster, and yet another Anglo-Saxon slaves mining lead under the guard of a Roman centurion. There is a touch of irony, surely, in the display showing a Victorian mother and daughter being guided round the cave a century ago.

Of the cave's natural features the most memorable is the 50-foot high hanging wall of the Great Rake, the scale for which is provided by the figure pushing a wagon-load of mineral.

Temple Mine

Location: Derbyshire
Temple Walk, Matlock Bath; just off the A6

Open:
Daily, Good Friday to 30 September, 11.00 am to 5.00 pm

Though the part of Temple Mine that is open today was first worked, for fluorspar, in the 1920s, the title to other parts goes back several hundred years, to a time when lead was actively mined as well. Most of the tour runs through passages with plenty of head room, but the first part is low and so safety helmets are provided. This first section is a level walk along a passage made large enough to allow ponies to haul out the ore. This was rare in Derbyshire, as most mine passages were too small to accommodate them.

The vein containing the fluorspar occurs at intervals in regularly shaped pockets just above the bed of rock known as the Lower Matlock Lava. After negotiating the entrance level, you turn off into a crosscut driven as late as the 1950s, by which time diesel trucks were being used for haulage instead of ponies. This crosscut leads to another pocket of ore, and from this an incline was cut upwards to meet another mine, but the connection was never made. Follow this incline up to about the halfway point (concrete steps have been added to make the going easier), where a passage has been cut to double back over the top of the last pocket of ore. This passage is followed to the conclusion of the tour at the point where deep and glutinous mud makes further progress impracticable.

Various items of pneumatic drilling equipment are displayed in the mine, and there is a tableau with life-size figures depicting miners in various stages of shot-firing. On the surface are specimens of mine machinery brought in from other mines.

Heights of Abraham

Location: Derbyshire
Signposted from the A6 in Matlock Bath

Open:
Daily, Easter to 30 September, 10.00 am to 6.00 pm
For winter openings, telephone Matlock 2365

Two centuries ago the southern slope of Masson Hill, above the village of Matlock Bath, was a waste scarred from hundreds of years of lead mining in and around the Nestus Mine, one of the oldest in Derbyshire. The Heights are traversed, from east to west, by one of Derbyshire's most important mineral veins – the Great Rake. In Roman times lead ore was dug easily on the surface of the hill, but later increased demand meant that the vein had to be followed underground.

In 1780, 30 acres of the hillside were formed into an estate named after the famous landmark at Quebec scaled by General Wolfe's troops twenty-one years earlier: the Heights of Abraham. The estate was developed as a pleasure garden with superb views of the Derwent valley. Development has intensified recently, and there are now a number of diversions including, as well as two cave systems, a Victorian prospect tower, a play area, a Tree Tops Visitors' Centre, and – the most spectacular addition – a cable car soaring above the railway, across the river and the A6, and up past Long Tor escarpment and through the woodland of the Heights.

Great Rutland Cavern and Nestus Mine

The main underground tour is through Great Rutland Cavern and Nestus Mine, once owned by the Dukes of Rutland. Known throughout its working life simply as Nestus Mine, it acquired its grander title when it was opened to the public in 1810, the miners having discovered the natural cavities some time earlier.

The system is entered through a 230-foot long passage which intersects layers of shale, limestone and clay. Halfway along, on the right, is a partly filled working that once connected with Lower Nestus – another area of the mine, now mostly flooded. The passage continues as a worked-out vein to the Roman Hall. As the passage begins to open out you can see, on the left-hand wall, an excellent cross-section of the mineral vein. A layer of basalt, which the miners called toadstone, caps several layers of minerals and ores common to this part of Derbyshire. From top to bottom they are: fluorspar, calcite, barytes, galena (lead ore) and zinc blende. On the roof above this vein are small deposits of the green copper mineral malachite. But the mine's riches do not end there; elsewhere are deposits of smithsonite (more commonly known as calamine), azurite, dogtooth spar and quartz.

A short walk now takes you to the Roman staircase, constructed by the miners from mining rubble to give access to the higher workings off the main hall. From here the way continues along a wide and high stope working (where the vein has been mined). Halfway up the stope, on the left, is an example of the plug-and-feathers used to mine the limestone. At the far end of the stope hangs a corfe – a seventeenth-century ore bucket in which the galena was taken up shafts to the surface some 100 feet above. The wooden stemple still wedged between the walls above the corfe is of the type commonly used as a climbing aid or platform for working the high, narrow rifts.

Opposite the corfe is the entrance to Nestor Grotto, the largest of the natural chambers, measuring 100 feet long, 33 feet wide and 42 feet high. A shaft in the roof of this chamber is still open to the surface, and the top can be located just behind the Visitors' Centre near the Masson Mine entrance. The chamber floor was used by the miners as a spoil tip and for washing ore with water from a nearby underground stream. On the

wall behind you are still signs of those early miners – pick marks, drill holes, and even soot stains from candles, their only form of lighting.

The tour now returns to the Roman hall for the conclusion, an audio-visual display recreating the mine during a working day in 1675.

Masson Cavern

This show-cave-cum-mine, which is viewed by the light of hurricane lamps carried by each visitor, lies behind the Visitors' Centre. The route follows just a short part of the many miles of passages in what is the largest accessible system in the county.

The entrance, a sloping passage some 150 feet long, follows the Great Rake mentioned earlier. At the end of this passage a chamber is entered, and the route goes to the right into a long series of workings under the old Nestus title. Passages off each side of the tourist route lead into a maze of lead and fluorspar workings, some of which were worked up to the 1950s. Formerly the route extended to a second and higher entrance, but fluorspar quarrying on Masson Hill restricts the present end to the top of a series of steps some 650 feet from the entrance. The return to the surface is back along the same route.

Poole's Cavern

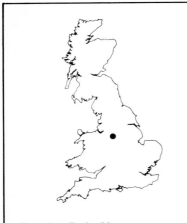

Location: Derbyshire
Buxton Country Park, off Green
Lane, Buxton

Open:
Daily, Easter to first week in
November, 10.00 am to 5.00 pm;
closed Wednesdays except from mid-
July to first week of September. At
other times by appointment;
telephone Buxton 6978

The cave takes its name from an outlaw
who, the story has it, lived in its
entrance in about 1440 – the gash in the
limestone protecting not only him but
his 'earnings' as well. Before the mid-
nineteenth century, visitors to the cave
simply wandered in at will with a scrap
of candle or a wooden torch. The
owners of the cottage near the entrance
latched on to the obvious opportunity
and, for a few pence, would guide
visitors through. The popularity of the
cave increased, even though the first 60
feet had to be negotiated by crawling.
Things went on to a more business-like
footing in 1854 when the landowner,
the Duke of Devonshire, became concerned at the amount of damage being
done in the cave. Free access had led a
certain number of people to regard the
cave's formations as fair game for
garden decorations, or whatever, and
the vandalism had increased with the
opening of the High Peak Railway and
the subsequent increase in the number
of visitors to Buxton. Poole's was to be
turned into a proper show cave, with
guided tours. The first custodian was
Frank Redfern, in whose family the
cave remained for well over a century.
Redfern's first job was to prepare a
proper pathway and install lighting.
Candles sufficed at first, but in 1859
seventeen gas-mantle cluster lights were

installed, and these, remarkably, remained in use until the cave closed in
1965, on the death of their owner.
Modern floodlamps were installed in
1976 and the cave was officially re-
opened, as part of the Buxton Country
Park, in the following year.

After only 60 feet, the narrow
entrance passage starts to widen into
Roman Chamber. This is no fanciful
name, for there is a continuing
archaeological dig here which you can
inspect from the pathway. So far it has
yielded over 2000 artefacts, including
sixty brooches. It is thought that this
may have been a Roman shrine. While
Neolithic man probably lived in the

cave entrance, the most important prehistoric period for Poole's Cavern was
the Iron Age. Recent evidence has
shown that Iron Age people occupied
and cooked food in the Roman
Chamber.

The next chamber, the Dome, is
about 50 feet high, and was probably
scooped out of the rock by the great
swirling torrents that the melting ice
produced at the end of the Great Ice
Age. The route threads its way through
the countless tons of boulders and other
debris washed in during this period.

In times of high water, you will now
begin to notice the unmistakable sound
of gushing water, for here is the source

of Derbyshire's river Wye. In the summer months the river is often dried up. When it is running, however, you now pass the point at which it vanishes into the boulders of the cave floor to run underground before resurging ¼ mile away at Wye Head. Here too is the Petrifying Well where the Victorians had a penchant for submerging things such as birds' nests. In time the calcite-rich water petrified them, producing another oddity for the mantelpiece.

A little further along the wide passage, at the prosaically named Constant Drip, is an example of how cave waters frequently give, then take away. Over thousands of years, the drip of water here first painstakingly built up a stalagmite, then, because of some change in the rate of flow or acidity of the water, drilled a hole through it. The extra white brilliance of the stalagmite nearby is a result of the water that formed it picking up, in comparatively recent times, an extra ration of lime from the old lime tips on the hillside above, a phenomenon which occurs again later in the cave.

Sadly, the largest stalactite in Poole's

— known by its old name of the Flitch of Bacon — hangs down from the middle of the roof in a slightly truncated version of its former self. Although nearly 6 feet long, a fair portion of it was knocked off, quite deliberately, at some time in earlier years.

From the Balcony is a splendid vista right back down the main passage for over 400 feet to the Roman Chamber. A little further in, though, round a slight bend, is the highlight of Poole's Cavern: Poached Egg Chamber. Here is a proliferation of many kinds of formation — straws, flowstone, stalactites and stalagmites — coloured in white, orange and blue-grey. This last coloration comes from manganese deposits, while the chamber takes its name from iron compounds tinting the tips of stalagmites a bright orange. In this chamber, too, the occasional very white flowstone overlay seems to stem from the old lime tips above.

Past more formations and you will come to the Mary Queen of Scots Pillar, a coloured stalagmite boss over 6 feet tall, supposedly marking the point at which the ill-fated queen curtailed her

Above The path skirts the pitted floor in this fine length of passage in Poole's Cavern.

Left Stalactite and stalagmite formation in Poole's Cavern.

visit to the cave in 1582. Now the passage twists back on itself, taking you into the final gallery. A descent of eight steps (there are only sixteen in the whole cave) brings you to the point where the river Wye emerges after its mile-long underground journey. There are plenty of formations here, the Grand Cascade — a flowstone mass flowing into finely ribbed pendants — being coloured orange and blue. After a journey of about 1000 feet from daylight, and at a depth of over 150 feet, you can look up to the huge pile of boulders which marks the present end of Poole's Cavern. But cavers being cavers, and boulders being (eventually!) moveable, we may well see the day when yet more passage is revealed.

Chatterley Whitfield Mining Museum

Location: Staffordshire
Just to the east of the A527 outside
Tunstall, about 5 miles north of
Stoke-on-Trent. Follow local
signposts

Open:
Tuesdays to Sundays and Bank
Holiday Mondays, from 9.30 am.
Last tour starts at 3.30 pm. Wear
stout shoes and warm, serviceable
clothing. Children under ten allowed
in all surface facilities but not
underground

Not so many years ago, the word
museum would evoke visions of long,
ill-lit rooms lined with dusty glass-cases.
While some museums still seem to
be presented as carefully preserved
exhibits themselves, in many others the
designers have been allowed to get to
work. Recently, too, numerous special-
interest museums have been set up,
often associated with just one aspect of
the work, crafts or way of life of the
past. The results of all this new thinking
are refreshing and exciting, as Chatterley Whitfield shows.

It is natural that the British should be
interested in coal. On it was founded the
nation's industrial power and wealth,
and, North Sea oil having apparently
already given of its best, coal looks set
again to play an important role as a fuel
source for a long time to come.

So a modern mining museum is of
particular interest, especially when it is
based on an actual mine, as is the case at
Chatterley Whitfield. Here, 700 feet
underground, we can see the varied
conditions under which coal was and is
mined, and the machinery and techniques that had to be developed over the
years to cope with the growing appetite
for this invaluable source of energy.

Based on Stoke-on-Trent, the triangular North Staffordshire coalfield
covers about 20 square miles, with

Chatterley Whitfield colliery in the
north-east. Started in 1860, the colliery
became the first in Britain to break the
significant barrier of 1 million tons
output in one year. Despite this record,
it was closed in 1976. In the face of
strong foreign competition, cost-effec-
tive production is a keynote of modern
mining, and it became more efficient
to recover the remaining reserves of
coal from Wolstanton colliery, 4 miles
away.

Before going underground, let us
loiter a moment on the surface. Visitors
who – despite the entreaties of this
book – cannot bring themselves to take
the drop into the dark (or children
under the age of ten, who are not
allowed down the mine) will find plenty
here to interest them, and are charged a
separate, reduced admission charge.

Towering above the stolid, no-frills
buildings is the gracefully tapering
chimney, over 200 feet high. This
splendid chimney has been the target of
a demolition crew, but only to lop off
the top 20 feet for safety reasons. Of the
numerous shafts sunk within the
colliery's boundaries, only four are now
readily recognizable by their tell-tale
winding headgear. The museum continues to run the Institute Shaft, the one
with the distinctive lattice girder head-
stock, to extract air from the mine and
as an alternative, emergency exit.
Visitors can examine the original steam
winding-engine which served Hesketh,
at nearly 2000 feet the deepest shaft,
now capped by concrete.

Other surface buildings include
various fan houses (which, in their
heyday, sucked 30 million cubic feet of
air from the mines every hour) and a
methane house where, with more than
just a nod to the maxim 'Waste not
want not', natural gas extracted from
the mine was used to heat boilers.

The latest 'surface' attraction is an
exhibition sited in the fandrift – the
ventilation system of the old colliery,
about 15 feet underground. This com-
plements the underground tour, providing appropriate sound effects,
models of pit ponies, examples of early
lighting, miners working at the coal-
face, and so on. There are no age
restrictions in this exhibition.

Interesting as the surface displays
are, the most fascinating aspect of the
museum is the mine proper. But before
leaving the daylight, you will be issued
with safety helmets, electric headlamps,
self-rescuers (emergency breathing

units) and riding checks – a simple
system of tokens which ensures that all
who go down the pit can be accounted
for on the return. Just as important is
the careful check for contraband; not
only matches and lighters, but anything
battery-powered – such as some
watches and cameras – which could
present a fire hazard.

While surface subsidence may have
to be tolerated in some areas around
mines, the one place that must be kept
stable is the colliery site itself. This is
achieved by leaving a shaft pillar, an
area of unworked coal, and it is in this
area that the underground part of the
museum is situated.

After he has checked everyone into
the cage at the top of the Winstanley
Shaft, the banksman signals to the
winding-engine operator using a coded
system of bells. The 700-foot descent
starts, taking about a minute in all. As
the cage descends, a counterbalanced
twin rises from the depths. This shaft
also serves another vital purpose as the
downward ventilation shaft for the
mine.

At the pit bottom you are met by the
onsetter whose job is like that of the
banksman up on the surface, and the
tour begins.

Displayed at the start are man-riding
trolleys like those in which the miners
travelled to their workfaces, as much as
4 miles distant, pulled by a locomotive
or – on steep sections – haulage ropes.

The first length of roadway (access
passage) ends where the main mine
workings have been sealed off from this
museum section by a number of walls.
The hydrant stationed here is for fire-
fighting. A short way back again from
here is the roadway you now take after
examining the workshop off to the
right. Once this contained a haulage
engine, and, being relatively draught-
free, it later proved a more comfortable
place for straightforward repair work
and meal breaks.

The tour now turns off to the right
and heads towards the Old Stable Road.
But before this there are several displays. The first, on the left at the junction where both old timber and modern
reinforced-concrete roof props are
used, is the top of the Holly Lane Main
Dip. Up this 1000-yard long incline coal
was hauled from the workface to where
it could be loaded into cages for the
straight lift to the surface. The Dip,
which was driven at the same angle as
the coal seam itself, is now also sealed
off.

Adjacent to where the Dip leads off, and on the same side, is a roadway which, being cut on the level across dipping strata, takes the name – peculiar to the region – of a crut. This, the Bell-ringer Crut, gave the miners access to the coal seam of that name; at its entrance is displayed a stage loader – machinery used to transfer coal from one conveyor belt to another, or from a belt to tubs.

Now you walk along the Old Stable Road where, until the 1930s, the pit ponies lived out their non-working hours 700 feet from sunlight, except for a brief excursion to the surface during the miners' annual holiday.

The stone-dust barriers across the roof of this roadway are a simple but effective precaution against the spread of an explosion in the mine. The shock wave which travels ahead of the igniting gases scatters the dust into the air, making it less inflammable.

Thorough ventilation of mines is necessary to provide a comfortable working atmosphere, of course. Even more important, it also removes naturally occurring explosive as well as poisonous gases. As already mentioned, the air is drawn in down Winstanley Shaft. It is circulated through all the passages and workings in a carefully controlled pattern before being drawn to the surface again through another shaft.

To make sure that the air does not simply follow the easiest and quickest direct route between the two shafts, a series of double air-lock doors keeps it channelled on its proper course. You turn off the Old Stable Road through the first of three such pairs of doors into an area where mining history has been unrolled chronologically and displayed in the most effective and memorable way: actually at the face of a coal seam.

The earliest coal mines were the simplest, but then the seams were only a few yards from the surface, or even at the surface where tilting rock strata brought the black band of fuel into the light of day. This first display shows such a primitive mine – the bell pit. The miners dug through the overlying stone until the seam was met, then dug out in all directions, giving the distinctive bell shape. When the stone roof had been undercut almost to the point of collapse, the whole working was abandoned and a fresh one started nearby.

This method became more and more inefficient as the deeper coal seams were worked. This was when the pillar-and-

Above Visitors leaving the cage at the pit bottom before the start of the tour.

stall system was evolved. The coal seam was dug out to its limits, pillars of coal being left behind to support the roof. Then the miners worked back to their starting-point, digging away the pillars and allowing the roof to collapse over the worked-out section.

The following display shows the next major development, the longwall technique, concentrating on its earliest form, in the late 19th century, when the mining was done entirely by hand.

The name itself hints at the chief characteristic of the technique. A coal face up to 1000 feet long was worked along its whole length, access for miners being via the roadways driven at each end. First the face was undercut, then the top layer was brought down when it was, in the miners' terms, 'ripe'. As the seam was mined, the worked-out section behind the miners was allowed to collapse at a (usually) controlled rate.

Chatterley Whitfield Mining Museum

Mechanization brought not only some lessening of the sheer slog of purely manual mining, but also much greater output. The next display shows a coal-cutter, which, working like a chain-saw, could undercut the seam with ease. It was driven by compressed air, the versatile power of which was also turned to a number of other mechanical aids.

The last of these displays brings the story right up to date with a demonstration of the super-efficient shearer/loader. This consists of a large-toothed disc which scythes sideways along the seam, removing on each pass a great swath of coal which falls on to a chain conveyor. Essential to this system are the hydraulic roof supports which can be lowered in sequence, advanced, and then jacked up to provide a solid support for the roof.

Rounding the corner from this last demonstration of the most modern mining technique, you pass to the side of a small area of rockfall, a sobering reminder of the constant risk involved

in mining, however sophisticated the technology of the day.

As you complete what is a loop within the tour, you pass examples of trolleys – the flat wagons that are the workhorses for transporting materials and equipment from one part of the mine to another.

Now you retrace your steps along the Old Stable Road, but turn left at the end to go through two air-lock doors which you passed earlier. The opening on the left-hand side immediately beyond these doors is where a haulage engine was housed. The great size of this engine can be guessed by the size of the room necessary to accommodate it. The engine drove an endless cable, at about half a steady walking speed, which stretched for 1 mile before looping round a giant pulley wheel. You can see here some of the clips by which trolleys and cars were fastened to this cable.

Ahead of you is the long main roadway leading to the Institute Shaft, which doubles as an emergency exit from the mine and as the extraction

Above Visitors at the coalface 700ft underground.

shaft for the ventilation system. Along this roadway, the roof of which is supported by both steel arches and – at the lower places – concrete girders, is a drilling rig used to tap pockets of methane gas. This keeps the gas released into a mine at a safer level, and gives the useful bonus of free fuel for boilers on the surface.

Just before the Institute Shaft, you turn right off the roadway on the last leg of the tour. About 1 million gallons of water enter the museum's passages each week, and this collects at the water lodge (on the right) where electric turbine pumps send it on its way into old workings.

Another sharp bend, through the final pair of doors, and you are once more at the foot of Winstanley Shaft, ready for the soaring ride up to daylight.

Ingleborough Cave

Location: North Yorkshire
By footpath from the A65, 1¼ miles
north of Clapham

Open:
Daily, 1 March to 31 October, 10.30
am to 5.30 pm. Tours start on the
half hour and last about 45 minutes
1 November to 28 February,
weekends, 10.30 am to 4.30 pm, or by
appointment, telephone Clapham 242

Gaping Ghyll, in the flanks of Ingleborough mountain, is regarded as the
grandfather of all northern potholes.
As well as an impressive network of
passages and shafts, it is best enjoyed by
cavers for its enormous main chamber
– a cavity large enough to swallow York
Minster, with a 365-foot-deep shaft
piercing through its great roof.

For more than a century one of the
greatest unbeaten challenges to northern cavers was to follow the course of
Gaping Ghyll's underground river. The
dream was of a through-trip from GG,
as it is known affectionately, to the
resurgence at Ingleborough Cave. The
actual forging of the link lay in the
hands of a small number of highly
experienced and dedicated cave-divers.
After decades of extending first this
passage, then that, and extending the
known limits of sump after sump, the
breakthrough was finally made in 1983.
The cavers' eventual success depended
considerably on some remarkably
accurate radio-location work, using
equipment underground (in both caves)
and also directly above on the surface.

This showed the cavers exactly where to
dig through boulders to make the connection.

Now at last Gaping Ghyll and
Ingleborough Cave can be claimed as
one system, even though only cave-
divers can make a full through-trip
from one to the other.

While entry into the 7-mile-long
system of Gaping Ghyll is difficult, and
for experienced cavers only (except at
certain Bank Holiday weekends when
local caving clubs erect a chair winch at
the main shaft), the initial ¼ mile of
passage in Ingleborough Cave has been
opened to the public.

The cave itself is situated in
Clapdale, a wooded valley running
roughly northwards from the village of
Clapham towards the blunt heights of
Ingleborough. The start of the path is

Below These isolated stalagmite columns in
Ingleborough Cave – aptly named the
'Skittles' – have grown from drip water
released by the small roof formations.

Ingleborough Cave

quite difficult to find, and so a little more detail than usual is given here. From the car park, cross the stream by the footbridge opposite the entrance to the car park, and then follow the road that runs upstream and parallel with the stream past the church to the entrance to the timber yard. Alternatively, you can take the parallel path through the wooded Ingleborough Estate Grounds, incorporating the Reginald Farrer Nature Trail, for which a small charge is made. Both ways join just before the cave. Allow 30 to 40 minutes, whichever route you take.

The entrance to Ingleborough Cave (sometimes also known as Clapham Cave) is a low, but wide arch in a small limestone cliff. The other entrance, close by but rather lower, is Clapham Beck Head Cave from which the Clapham Beck now issues. Where the entrance passage of Ingleborough Cave narrows to a height and width of about 8 feet, a wall and gate have been built. The lime deposits you can see at this point are known as tufa.

After only 70 feet you are at the old limit of the cave, in terms of human accessibility, for a 5-foot-high calcite

barrier once spanned the passage here, damming a huge lake of water which came to within several inches of the roof. In 1837 the owner of the Ingleborough estate had the barrier breached; the water flooded out, revealing the cave as we see it today. Beyond the barrier, the broken ends of which still stand on each side of the path, the waterline of the old lake is quite clearly discernible at about shoulder height on each side. At the Inverted Forest, stalactites grew downwards into the lake surface and, instead of continuing their normal growth to shapely points, grew bulbous ends by gathering tufa from the water. The large stalactite a little further on shows two waterlines, indicating high- and low-water conditions.

Now the passage ascends slightly and opens up into Eldon Hall. To the right is the Mushroom Bed, a large stalagmite formation which grew outwards on meeting the old water level, giving it a mushroom shape. The rippling effect on its surface is caused by the growth of tiny gour pools, each a miniature dam often containing crystals. Here also in Eldon Hall is a group of helictites, a

Above The Curtain Range in Ingleborough Cave.

comparatively rare formation, growing outwards and even upwards.

The original explorers found yet another barrier past this point, most of which still stands, but a breach was made easily to drain the dammed water beyond. This second lake reached a height which is clearly indicated by the base of the Beehive, the next large formation on the left-hand wall. The 5-inch stalactites beneath have grown in only the century and a half since the lake was drained, which rather puts paid to the old rule of thumb many cave guides use, that stalactites grow 1 inch every 100 years. The walls below the old waterline are encrusted with knobbly lime deposits which crystallized from the mineral-rich water. The impressive 5-foot stalactite here is the Sword of Damocles.

Now the cave passage increases in size, and the roof is decked with a number of stalactites. Two pools, which once would have presented a damp and daunting obstacle, are now spanned by a bridge which takes you dry-shod and comfortably into Pillar Hall. The big stalagmite here is the Jockey's Cap – some 3 feet high and 10 feet in circumference. Measurements have been made of this formation since 1839, but sadly the different observers over the decades have not followed any standardized system of measuring, so the results are not as valuable as they could have been. Measurements taken in the late 1960s indicated (no researcher will be pinned to a firmer statement) that the Jockey's Cap had grown only ½ inch higher over the previous sixty-three years.

To the right are two stolid columns given the name the Elephant's Legs; opposite is a low section with numerous delicate straw stalactites and, further along, the Rippling Cascade. This whole chamber takes its name from a decidedly elegant column, over 8 feet tall, called very simply The Pillar. More ambitious nomenclature has been called into service with the christening of the 20-foot-deep hole which follows: The Abyss. This takes most of the water from this part of the cave, both at normal and flood level. The passage beyond can be followed by cavers for only a few yards before sumping.

Now an even larger passage, 15 to 20 feet high and wide, takes us into the Curtain Range. Two large stalactites on the left, the Shower Bath, have been deeply corroded by an increased flow of water down them, and shortly beyond are the Curtain Stalactites, draping down the wall. The passage in this section shows a number of features of water erosion, particularly in the form of the shell-like depressions known as scallops. The inlet stream which comes in here is still eroding a channel across the floor.

The yellow or white patches which you see on much of the passage wall from here on are mainly colonies of a primitive form of plant life, wall fungus; a few others are splashes from old candles. The roof now lowers, but do not miss the area of sand and gravel on the right which was bonded together by calcite, after which the loose material below was washed away. After stooping for a few feet, you can again stand up where the roof rises into a Gothic arch, the effect being enhanced by calcite curtains on each side. Behind the curtains lie two stalagmite bosses, the Ladies' Cushions. But walking to the left, along the Gothic Arch passage, you enter a new section of cave similar to the one before the stoop. Stalactites

Top The great stalagmite bulk of the Mushroom Bed in Eldon Hall.

Left Visitors' viewpoint – overlooking the pellucid Crystal Pool.

and stalagmites adorn each side, and one group has grown to form a miniature dam, holding back a small lake.

Now you are in the Long Gallery, and again you can see, on the right, patches of gravel cemented together and then undercut by the water. After the Coffee Pot stalagmite on the right of the path, and another small, low inlet on the left with a stalagmite cascade beneath, the passage divides for a few yards. The path goes to the right, but a look into the left-hand branch reveals the group of stalagmites which have given name to this passage: Skittle Alley.

At this point, ¼ mile from the entrance, the path – and the tour – finish. But there is one more surprise, for as you turn back another branch passage comes into view on the right, and a shallow pool faithfully reflects the roof stalactites in the Pool of Reflections.

As you retrace your steps to daylight past the various barriers, bear in mind that their removal marked the first step in what has been one of the most exciting stories of underground exploration in Britain. At the point where you turned round, ¼ mile in, another 2¼ miles of passage lay in front. At first the going, along Cellar Gallery (the continuation visible from the end of the path), is not too demanding, although only 3 to 4 feet high. But in the further reaches cavers have to crawl through the stream, turning their heads to one side to breathe in the 6-inch air-gap between water and roof. Eventually, series of sumps are met, both upstream and downstream towards the resurgence of Clapham Beck.

Such are the hazards and obstacles that confront any caver hoping to make the link with Gaping Ghyll, the main entrance shaft of which lies 1 mile to the north of the entrance to Ingleborough Cave.

White Scar Caves

Location: North Yorkshire
1 mile north-east of Ingleton on the
B6255

Open:
Daily, Easter to 31 October, 10.00 am
to 4.00 pm
In winter on request; telephone
Ingleton 41244

The entrance to this show cave nestles by the side of the determinedly straight road along the bottom of one of Yorkshire's most magnificent dales: Chapel-le-Dale. Behind the low white building which marks the entrance are the bottom-most flanks of Ingleborough, whose broad summit lies about 2 miles away. Many English valleys are modest, cloaking their sides with soil and vegetation. Chapel-le-Dale is strikingly different, for here you can see, rising for hundreds of feet, the bands of the Great Scar Limestone; the earth's structure is laid bare on a spectacular scale. In this dale it is easy to understand why the great limestone masses of Yorkshire house so many of the country's greatest cave systems. White Scar Caves give you a chance to peek into that great underworld and its unique character.

And it was a caver who discovered the system in 1923. Christopher Long (then an undergraduate at Cambridge) spotted the entrance from the other side of the dale and immediately crossed over to investigate. Though on his own, excitement at the find overcame any caution over solo caving – something rarely undertaken, even these days, in known caves, let alone in unexplored ones. He crawled in, managing to reach the Second Waterfall on that trip, about 1000 feet from daylight.

Long's route took him through many low and wet passages, quite out of the question for non-caving visitors. But the discovery of the active stream passage deep within the hillside proved sufficient incentive to the owners to blast a walking-size tunnel, chiefly along the course of the low natural passage.

At one point, however, about 20 yards in from the entrance, part of the original passage can still be seen, curving off to the right to rejoin the tunnel after a short way. From the comfort of the illuminated tunnel, demanding no more of the visitor than

Below More than 3 miles of wild cave can be explored by the caver in White Scar.

the wearing of a jacket in summer, the sight of that low crawl twisting off into the darkness provides a vivid picture of what Long's solo exploration demanded of him. Some it horrifies, others it fascinates. Certainly in the author – when a schoolboy visiting his first cave – it laid a seed of curiosity that grew and provoked over the years until he started his own caving career. But caving then was a far cry from today's wetsuit-clad and electrically-illuminated 'luxury'. Long made his exploration by the light of candles stuck on the brim of his hat: these were extinguished at one particularly low point where he had to push through the water on his back, his mouth to the ceiling to reach the small airspace.

Passing from the tunnelled section into the natural cave, you now hear the sound which must have thrilled Long: the roar of a cave stream. This is the First Waterfall, the vigorous stream cascading over the wall into a pool under a canopy of old formations. From here the water takes to a very low passage – passable by cavers only in dry conditions – and emerges north of the cave entrance.

The passage beyond the small hole at the top of the waterfall is your objective, but a blasted tunnel spares you the chilly climb and crawl which faced Long. Now follows natural passage, opening suddenly into a tall chamber, down which plummets the Second Waterfall, the noise reverberating off the heavily calcited walls.

Now the passage is beautifully decorated with all manner of formations – stalactites, delicious tiny gour pools, and swelling flowstone walls, the largest section of which swells fatly into the passage, earning it the name Buddha. Colours abound here, from bright reds and yellows to black and subtle greys.

From the Grotto, with its grouping of white stalagmites, the show cave progresses for a final 150 feet or so to end at the Barrier. Though one day access might be possible for visitors to the mighty chambers and formations beyond, this marks the point where the caving specialists take over, with a further 3½ miles of cave before them.

Below Straw Chamber in White Scar.

Stump Cross Caverns

Location: North Yorkshire
6 miles east of Grassington on the
B6265

Open:
Daily, 1 April to 31 October, 10.00
am to 5.30 pm
1 November to 31 March, weekends
only, 11.00 am to 4.00 pm

The accidental discovery of natural caverns by miners or quarrymen has been, and continues to be, a fairly frequent occurrence in most limestone regions of Britain. It would be unfair to a few to be too dogmatic, but by and large it does seem that, certainly in the case of miners, in the past they had more awe and respect for any particularly well-decorated caves they chanced upon. Perhaps today's commercial pressures are greater; perhaps today's valuation of such natural treasures is different. Whatever, there are more than a few instances in which the old miners left a natural cave in as undisturbed a state as possible.

Such is the case at Stump Cross Caverns. This cave was discovered in 1858 when miners broke into it by chance while digging for lead ore. They found a system so beautifully decorated that, although they were unsuccessful in their quest for lead, they left a permanent entrance in the form of a small shaft.

Nowadays visitors enter via stairs down a passage dug at a later date. At the foot of the stairs, the path leads off to the left. The partially blocked entrance after a short distance on the left is the entrance to the lower series, accessible only to cavers, and the site in 1963 of Geoffrey Workman's then

record-breaking stay underground, alone, for 105 days.

About 70 feet from the foot of the stairs, on the left, is a section known as the Butcher's Shop, almost closed off by the rich growths of stalactites and stalagmites, many of them coloured red from iron in the limestone. Close by is a crevice leading up into the roof with a series of calcite cascades. Just beyond are the Twins – two stalactites sharing the same stalagmite boss.

Now comes a junction where you walk to the right (the other passages are visited on the return journey, a treat worth waiting for). Some 40 feet beyond the junction, in a section of the passage which sports a finely domed roof, solution pockets are visible, dating from when the entire passage was water-filled. The present water level is about 150 feet below, the stream constantly seeking weaknesses in the rock to exploit and widen into cave.

Where the passage does a dog-leg bend to the right and then the left, look out for the small notice announcing the Jewel Box. Peer through a small hole in the wall and you will see a miniature cave beyond, with small crystal-filled pools. Beyond the dog-leg is the Snowdrift, a deposit of pure calcite, and, after a section of small passage and a sharp bend to the left, Jacob's Well – a gour pool fed by water drips, surrounded by sparkling white calcite.

Perched on a rocky slope on the left is the Hawk, a stalagmite guarding one of the treasures of Stump Cross Caverns: the Sentinel. This superb column, nearly 10 feet high, falls from the roof as an exquisite stalactite before joining the stalagmite below, the base of which spreads out, mountain-like, to cover almost the entire passage floor. In turn the Sentinel – reckoned by some to be at least 200,000 years old – guards the entrance to a splendid series of chambers.

In the first of these, the Chamber of Pillars, columns are predominant, but

to one side a large rock has been transformed by calcite flow into the Cradle. A short way ahead, on the left, is the Sandcastle Grotto, where there are several active gour pools and sandcastle-shaped stalagmites. Now the passage bends sharply to the right into the final and largest chamber, the Cathedral. Beneath a roof festooned with stalactite straws perches the Organ and, on the final boulder choke which marks the end, the stalagmite column whose shape led to its name, the Wedding Cake.

Now you retrace your steps, but only as far as that first large junction, for the tour is by no means finished. At the junction, turn right into Wolverine Cave. The decorations start immediately, for in the rift above hang calcite curtains. This part of the system is the latest to be added to the visitor's tour, and the floor here has been excavated to provide easy access. From the corner ahead you have your first sight of the many fine formations which decorate this passage, with flowstone spreading

Above Reindeer Cave, Stump Cross Caverns.

Left The Sentinel, Stump Cross Caverns.

over both walls and stalactites over the roof. Now a mass of gour pools appear, and a short flight of steps leading down. In this section excavations revealed bones of wolverine, wolf, bison and reindeer. The wolverine skull and jaws created considerable scientific interest, and are now held by the Natural History Museum in London. The Museum presented the management of Stump Cross Caverns with replicas, set in the very deposits in which they were found, and these are now on show at the cave.

The walk along Wolverine Cave takes you through a wealth of formations of every kind until, at the end, they grow so profusely that some would have had to be destroyed to allow further access. Fortunately, this has not been done.

SCOTLAND

There are few major underground sites open to the public in Scotland, mainly because of the relative paucity of its limestone caving areas. Caves there are – the classic area being the parish of Assynt in the far north of the Highland region – but the great majority are short, and have a deserved reputation for difficulty, and a high degree of dampness. Despite this sporting element, relatively few English cavers make the journey north in pursuit of their pastime.

Sea caves abound along many stretches of Scotland's lengthy coastline, however. There are many on the Isle of Arran, and a number of others are scattered throughout the country.

While lead was never mined in Scotland as extensively as elsewhere in Great Britain, the chief mines at Wanlockhead and Leadhills were worked (for their gold yield as well) from the 17th to the early 20th century. At Wanlockhead, part of the Museum of Scottish Lead Mining, a mine level can be visited.

The tour of the Cruachan Hydro-electric Power Station in Strathclyde provides a completely modern contrast. An electric bus takes you into the heart of Ben Cruachan to an observation platform overlooking the huge underground halls of machinery.

Those with a taste for the spectacular should find an excursion to Smoo Cave amply repaid by its great entrance, opening up at beach level and easily penetrable for several hundred feet.

Cruachan Hydro-Electric Power Station

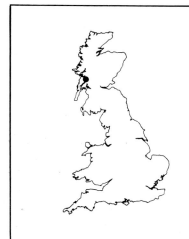

Location: Strathclyde
On A85, 14 miles east of Oban

Open:
Daily, Easter to 31 October, 9.00 am to 4.30 pm. In mid-season, arrive at Cruachan by 12.30 pm

Top Tunnel leading to underground power station.

Above Interior of the underground power station with visitors' viewing gallery and mural depicting the story of Cruachan.

One of the problems of electricity supply is that it is in high demand only at certain times of the day. This would not present too much of a problem if electricity could be generated at the drop of a switch. But it is not profitable to work the giant power-stations generating from fossil fuels full-out only in the peak periods. What are required, ideally, are huge batteries that can be charged throughout the slack periods, then discharged into the national grid when demand is high.

And huge batteries there are, in the form of pumped-storage power stations. At Cruachan, the first large-scale development of its kind in Scotland, the public can visit the very core of the mountainside which plays a key part in the whole working.

The engineering aspects of Cruachan are on an enormous scale, and yet nearly all of it is subterranean. From the administration block on the banks of Loch Awe, an electric bus takes you nearly ¾ mile straight into the side of Ben Cruachan. Then it is a short walk to the observation gallery overlooking the great machine hall − blasted out of the solid rock, and large enough to contain a seven-storey building covering the area of a full-size football pitch.

The four giant turbine generators produce about 450 million units of electricity each year, the water to power them falling through concrete-lined shafts nearly 1200 feet from a reservoir in a dammed corrie on Ben Cruachan. When demand for electricity is low, the turbine generators switch to their alternative role of pumps, returning water from Loch Awe up through the tunnels and shafts back to the reservoir, building up pressure ready for the next period of production. The power used to drive the generators as pumps comes from the other power stations that would otherwise be working at a low and inefficient off-peak level.

For all the size of the turbines/pumps (the rotating parts alone of each weighs 250 tonnes), they are remarkably responsive. From a standing start they can be operating as generators within two minutes, or, if spun by compressed air to start, within one minute.

Scotland may not have the great hydro-electric potential of, for example, Norway, but it does have mountains. And under Ben Cruachan, ingenious and spectacular use has been made of one to help meet Britain's enormous appetite for electricity.

Smoo Cave

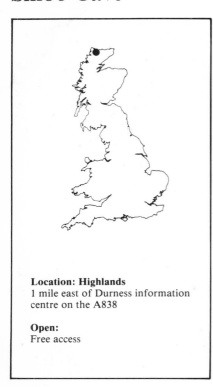

Location: Highlands
1 mile east of Durness information
centre on the A838

Open:
Free access

This cave, regarded as one of Scotland's grandest natural features, is reached easily along a footpath leading down to the beach from a small car park. Lying only a short distance above the high-water level, the spectacular entrance to Smoo measures about 120 feet across and rises to a height of 60 feet. Beyond is a huge rectangular chamber, going back over 200 feet into the cliff, its roof pierced by small blow holes.

This is basically a sea cave, as witnessed by the blow holes and other features. But the two small streams that unite inside the main chamber and flow out of the entrance down to the sea testify to some conventional inland cave formation as well. The left-hand stream flows from the bottom of a great flowstone cascade and can be followed by cavers to a tiny chamber with no further continuation. The right-hand stream has its source in a deep pool beneath a natural rock bridge. Beyond lies another large cavity, Lake Chamber, which cavers can reach by climbing over the bridge. The waterfall from another entrance, Allt Smoo, pours into the lake.

Right The view from inside Smoo Cave.

164

Museum of Scottish Lead Mining, Loch Nell Mine

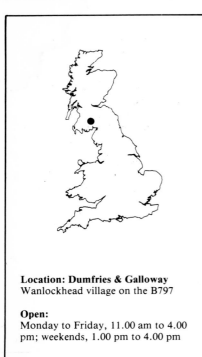

Location: Dumfries & Galloway
Wanlockhead village on the B797

Open:
Monday to Friday, 11.00 am to 4.00
pm; weekends, 1.00 pm to 4.00 pm

When Dorothy Wordsworth, sister of
the famous poet, visited the area in 1803
she described it as 'a valley which nature
has kept for herself'. Yet had she gone a
little further, she would have come to
the village of Wanlockhead, and a
rather different picture. At an altitude
of some 1400 feet, this is one of the
highest villages in Scotland, set in
isolation amid high, heather-covered
moorland whose slopes are marked by
old shaft heads, mine spoil tips, and
barren patches of ground where the
residue from smelting was dumped: the
scars left by a land worked for lead.

While never a major industry in
Scotland, lead mining has had a continuous history at the mines of
Wanlockhead and nearby Leadhills
from 1680 until the middle of this
century. Gold has been mined here,
too, and the 300 miners employed by
the Master of the Mint to Elizabeth I are
said to have obtained gold valued at
over £100,000. The description given to
the area of 'God's treasure-house in
Scotland' hardly seems all that fanciful
when one considers that the largest
piece of gold found here, now lodged
with the British Museum, weighed
between 4 and 5 ounces. Even as late as
the last century it was customary for
lead miners to mine enough gold to
make a ring for their brides.

The establishment and growth of the
village of Wanlockhead paralleled that
of mining activity in the immediate
area, and with the establishment of the
open-air Museum of Scottish Lead
Mining it has become an integral part of
the museum's mining tour. The trail
starts at the museum building, once a
miner's cottage with two rooms below
and very small attics. Close by is the
miners' library, established in 1756 by
men of the village for 'our mutual
improvement'. The restoration of this
library was one of the Museum Trust's
first tasks. The trail passes close by the
schoolhouse. When this was first built
– in the 1850s – the salaries of the
schoolteacher and two assistants were
paid by the Duke of Buccleuch and
Queensbury. As attendance at the
school had declined from its original
136 pupils to only 11 in 1977, it was
closed, but today serves in a new role as
a community hall and field studies
centre.

Loch Nell Mine is encountered in the
early stages of the trail, virtually within
the village itself. For the miners of
Wanlockhead, work was inextricably
woven into social life. Some 400 yards
of the mine have been opened up for
visitors to gain a first-hand impression
of what was involved in lead mining
here.

Loch Nell was started shortly after
1710 by a company that had found rich
lead ore at the northern end of Cove
Vein. They made two trial tunnels at the
southern end of the vein, in the middle
of the village: Williamson's Level,
driven into Dod Hill from the side of the
Wanlock burn, and the High Level, 150
feet further up the hill. But the returns
were poor and the two levels, having
been driven about 100 yards, were
abandoned.

The Wanlockhead lease was taken
over in 1756 by Ronald Crawford &
Company, and mining of the two levels
was resumed. The lead yield of the vein
improved, but poor ventilation put
another stop to work. For a time
Williamson's was used as a reservoir for
water used to turn waterwheel pumps at
the adjacent mine of Straitsteps. A
crosscut was made from Straitsteps in
1793, intercepting the Cove Vein 90 feet
below Williamson's Level. The two
original levels were re-opened once
again, and a third was driven from the
Straitsteps crosscut (Thomson's Level).
Some 400 yards in, two shafts were sunk
to connect the three levels, considerably
easing both access and ventilation. Both

upper levels were extended for several
hundred yards, lead ore being mined
successfully above both, and, as it was
extended, ore extraction was made
possible above Thomson's Level.

By the 1830s shafts and drifts had
been opened below Thomson's Level to
a depth of 100 feet, but with only hand-
pumps to remove the constant ingress
of water the viability of the mine was
marginal, and the workings were again
abandoned. In the mid-1840s a water-
pressure pump was installed in
Thomson's Level, supplied with water
piped in along Williamson's Level,
which also carried a small railway.
These steps enabled the workings to be
lengthened and taken down to 800 feet
below Thomson's Level. Water became
a problem again only twenty years later,
and the mine was finally closed.

Loch Nell was re-opened for a
geological survey in the 1950s, but
mining did not follow, and it was closed
yet again until the museum took it over
in 1979. The visit will take you some 400
yards along Williamson's Level, to just
beyond the connecting shafts between
the different levels. In the first 20 yards
the level is crosscut to the vein, and the
original timbering through the loose
shale and rock is preserved. The level
follows the course of the vein, each side
displaying the different characteristics
of both sides of the mineralized fault.
Most of the level, the roof of which is
blackened by candle smoke, was driven
by hammer and wedge, but occasional
small drill holes indicate the use of gunpowder too. The guided tour ends, just
past the 150-foot shaft up to High Level
and the 90-foot shaft down to
Thomson's, with a display of life-size
figures of miners demonstrating the different methods used to win the lead ore.

Back on the surface, the trail leads to
a number of interesting surface workings, among them the Wanlockhead
beam engine, the Pates Knowes Smelt
Mill, the surface remains of the Bay
Mine, and the waterwheel pit, whose
wheel was connected to the pumps in
the Bay Mine shaft by rods.

Places to Visit

The South-West

Buckfastleigh

William Pengelly Cave Studies Centre, Higher Kiln Quarry, Russets Lane, Buckfastleigh, Devon. Established in 1962, the centre operates as a focal point for the study of caves by research and through educational programmes, with a strong emphasis on cave conservation. The disused quarry in which the centre is located contains several caves, including Joint-Mitnor Cave with fossil bones dating back some 100,000 years. The centre is open to the public on selected days each summer; information from W. Joint, 93 Oaklands Park, Buckfastleigh, Devon.

Camborne

Camborne School of Mines Geological Museum, Pool, Redruth, Cornwall. Monday to Friday, 9.00am to 4.30pm. Holds a collection of minerals and ores.

Cornish Engines, East Pool Mine, Pool, Redruth, Cornwall. Daily, April to October, 11.00am to 5.30pm. This National Trust site displays two large mining-engines, each in its own housing; a 30-inch rotative beam winding engine, dating to 1887, and a 90-inch beam pumping engine of 1892.

Cinderford

Dean Heritage Museum, Camp Hill, Soudley, Cinderford, Gloucestershire. Daily, 10.00am to 6.00pm (July and August 10.00 am to 8.00 pm). The overall theme is Man in the Royal Forest of Dean; mining in the Forest of coal, iron and stone is shown, as well as the Freeminer at work and home.

Torquay

Torquay Natural History Society Museum, 529 Babbacombe Road, Torquay, Devon. Sunday to Friday, 10.00am to 4.45pm (also Saturday, March to October). Includes displays on Kents Cavern and other caves.

Wells

Wells Museum, Cathedral Green, Wells, Somerset. April to September, Monday to Saturday, 11.00am to 5.00pm; October to March, 2.00pm to 4.00pm (also Sunday, 2.30pm to 5.30pm, June to September). Contains displays of prehistoric finds from Mendip caves as well as local fossils and minerals.

Southern England

London

The Geological Museum, Exhibition Road, South Kensington, London. Monday to Saturday, 10.00am to 6.00pm; Sunday, 2.30pm to 6.00pm. Very extensive displays covering all aspects of geology.

Portsmouth

Fort Widley, Portsdown Hill Road, Portsmouth, Hampshire. Daily, April to September, 1.30pm to 5.30pm. One of the forts built by Palmerston in the 1860s under the threat of a French invasion; includes many underground tunnels.

Wales and the Borders

Aberystwyth

Llywernog Silver-Lead Mine, Ponterwyd, near Aberystwyth, Dyfed (on A44). Daily, Easter to 31 August, 10.00am to 6.00pm; September and October, 10.00am to 5.00pm.

This site has been preserved and restored as a typical example of a mid-19th century, water-powered metalliferous mine, though silver and lead were in fact mined here from about 1740 to 1914. The self-guided tour includes an excursion underground into Balcombe's Level (a prospecting cross-cut level dating from about 1790) for 70 yards. The electrically lit level, which is timbered for the first 12 feet, intersects the main ore lode, and there is a winze (a descending underground shaft). On the surface several waterwheels have been installed to show how water provided the mine's motive power, and above the main engine shaft, which was driven to a depth of 432 feet, rises the headframe. Well away from the main structures is the gunpowder magazine, built on the Cornish beehive pattern. In the restored rock-crushing house is an enormous Cornish crusher, and the jigger shed houses a three-compartment machine used to wash the heavy lead ore (with its silver content) from the waste rock. Close by are two round buddle pits in which were recovered the fine ore particles which had passed through the jigger. In the old ore-dressing shed is an audio-visual display, while the main museum building contains a 'California of Wales' exhibition as well as underground reconstructions.

Crynant

Cefn Coed Coal & Steam Centre, Blaenant Colliery, Crynant, Neath, West Glamorgan (on A4109). April to October, daily, 11.00am to 5.15pm; November to March, only parties of 10 or more by arrangement.

Situated next to a working colliery, this centre, opened in 1980, houses a number of carefully preserved mining machines — including a huge steam winding-engine — and various displays which give an introduction to the history of coal mining in general and in the Dulais Valley in particular. A 100-foot-long simulated mining gallery has been constructed a few feet underground to convey something of the atmosphere of work underground.

Dolgellau

Maesgwm Visitor Centre, Ganllwyd, Dolgellau, Gwynedd. (Follow the 'Forest Centre' signs on the A470 about 7 miles north of Dolgellau.) April to September, daily, 10.00am to 4.30pm.

As well as displays on the local geology there are two gold separation pans from a local gold mine. The crushed ore was ground to a fine powder by rollers in these pans, the particles of gold sinking to amalgamate with a layer of mercury at the bottom.

Llanberis

Welsh Slate Museum, Llanberis, Gwynedd. Easter Saturday to 30 April, 9.30am to 5.30pm; May to September, daily, 9.30am to 6.30pm. Displays of equipment and machinery connected with local slate-quarrying and -mining.

Port Talbot

Welsh Miners Museum, Afan Argoed Country Park, Cynonville, Port Talbot, West Glamorgan. April to October, daily, 10.30am to 6.30pm; November to March, Saturday and Sunday only, 12.00pm to 5.00pm.

Built in a once busy mining valley, the museum tells the story not only of the coal miner's work underground but also of his home life. The first display is a reconstruction of a typical domestic scene. The centrepiece of the circular tour comprises several underground simulations with pit props and roof supports to add to the atmosphere.

Places to Visit

Northern England

Batley

Bagshaw Museum, Wilton Park, Batley, West Yorkshire. Monday to Saturday, 10.00am to 5.00pm, Sunday 1.00pm to 5.00pm. Displays on the history of coal mining.

Beamish

North of England Open Air Museum, Beamish, near Chester-le-Street, County Durham. 19 April to mid-September, daily, 10.00am to 6.00pm; mid-September to mid-April, Tuesday to Sunday, 10.00am to 5.00pm. The museum, which covers 200 acres, is devoted to portraying aspects of northern life at about the turn of the century. Coal mining is included, with displays of a drift mine and furnished pit cottages.

Buxton

Buxton Museum & Art Gallery, Terrace Road, Buxton, Derbyshire. Extensive collections associated with the wealth of local caves as well as specimens and finished ornaments of Blue John, the mineral unique to the area.

Creswell

Creswell Crags & Visitor Centre, Crags Road, Welbeck, Worksop, Nottinghamshire (on B6042 between A616 and A60, 1 mile east of Creswell). June to September, centre Tuesday to Saturday 10.00am to 5.00pm, Sunday 10.00am to 7.00pm, crags daily; October, centre Tuesday to Sunday 10.00am to 4.30pm, crags daily; November and December, crags and centre, Sunday 10.00am to 4.30pm.

The crags are in a narrow limestone gorge, both sides of which contain a number of caves found to have been occupied by men and animals 70,000 years ago. Note that public access to these caves is prohibited. The visitor centre at the start of the trail has an exhibition and audio-visual display telling the story of prehistoric life in the caves. The five main caves are situated just above the present river level, but there is another layer of smaller caves higher up the crags.

Dudley

Dudley Museum & Art Gallery, St James's Road, Dudley, West Midlands. Monday to Saturday, 10.00am to 5.00pm. Has a fine geological gallery, with the emphasis on local coal measure fossils and limestones.

The Black Country Museum, Tipton Road, Dudley, West Midlands. April to December, Sunday to Friday, 10.00am to 5.00pm. Includes demonstrations of steam machinery excavating a mine shaft. The museum is built around the canal basin at the northern end of Dudley Canal Tunnel, into which trips are run from the museum.

Haughton

National Mining Museum, Lound Hall, Haughton, near Bothamsall, Retford, Nottinghamshire (next to Bevercotes Colliery off B6387). April to September, Tuesday to Saturday, 10.00am to 5.30pm (first Sunday of month, 2.00pm to 5.30pm); October to March, Tuesday to Saturday, 10.00am to 4.00pm (first Sunday of month, 2.00pm to 4.00pm). Many exhibits portraying the coal mining industry including headstocks, a canal boat, shunting locomotives, and many mining hand and power tools. The Ellistown pumping engine (and sometimes a steam crane) is brought 'on steam' on Saturdays.

Matlock Bath

Peak District Mining Museum, The Pavilion, Matlock Bath, Derbyshire. Daily, 11.00am to 4.00pm.

A large staging has been built in the museum's main hall dividing it horizontally. Below are reproductions of various aspects of lead mines. Above, displays show the surface features of mining and smelting of lead, the main industry in the Peak District for nearly 2000 years. Visitors can scramble down or along the shafts and levels built in this section to gain at least some idea of how the miners got to and from their places of work. The centrepiece of the museum is an enormous water-pressure engine designed by the Cornishman Richard Trevithick.

The museum also organizes conducted tours round the surface remains of Magpie Lead Mine in the Peak District. There are a number of buildings and other items of mining interest.

Newcastle-upon-Tyne

University of Newcastle upon Tyne, Museum of the Department of Mining Engineering, Queen Victoria Road, Newcastle upon Tyne. Monday to Friday 9.00am to 5.00pm. The department's museum has an extensive collection of mining safety lamps and other exhibits showing various aspects of the history of mining.

Northwich

Salt Museum, London Road, Northwich, Cheshire. Tuesday to Sunday, 2.00pm to 5.00pm (July and August also 10.00am to 1.00pm). Displays, models and reconstructions show the history of the salt mining industry in Cheshire from the Romans to today.

Salford

Salford Museum of Mining, Buile Hill Park, Eccles Old Road, Salford, Greater Manchester. Monday to Friday, 10.00am to 12.30pm, 1.30pm to 5.00pm; Sunday 2.00pm to 5.00pm.

Designed by the architect of the Houses of Parliament, the Georgian house which contains this excellent museum is the most unlikely setting for the two coal mines reproduced within its walls in a very convincing fashion — even down to thick layers of coal and stone dust.

Buile Hill No. 1 Drift Mine occupies the cellars, and recreates a typical small mine of this kind in the 1930s. A drift mine is one that operates where the coal is very close to the surface, the drift being driven down along the sloping seam without the need for deep shafts. Some of the remarkably life-like reconstructions encountered are a pit yard, blacksmith's forge, fan room, manager's office and pit head baths.

On the ground floor, Buile Hill No. 1 Pit has three main display areas. In the first, the visitor passes through the pit shaft cage to a typical hand-worked coal-face of the 1840s. Through a ventilation door and you jump a century forward, for here an undercutter is ripping along a longwall coal face. Round a corner is a rock face prepared for shot firing, and past this you make your way out along a small roadway line with herringbone-pattern pit props past the mine deputy's cabin.

The museum's first floor is devoted to displays of photographs and exhibits which tell the whole story of mining and that of the miner too.

Glossary

ABSEILING: Technique of sliding down a fixed rope in a controlled manner, usually using a special friction device (descender).

ACETYLENE LAMP: Lamp commonly used in caving in which water dripping on to calcium carbide generates acetylene gas, which is burnt in a jet.

ACTIVE CAVE/PASSAGE: One in which a stream still runs.

ADIT: (mining) Nearly level entrance, often used for drainage.

AVEN: Vertical hole in the roof of a cave, sometimes leading to higher passages.

BED: One of the strata layers in sedimentary rocks.

BEDDING PLANE: The plane separating two beds of rock; to a caver, a very low wide passage in such a plane.

BELAY: Natural or artificial anchor-point for ropes or ladders; (verb) the act of attachment to an anchor-point, or to safeguard a climbing caver by means of a lifeline.

BELL PIT: Early form of coal mine in which the shaft bottom was belled-out to work the seam (qv).

BLUE JOHN: Form of decorative fluorspar found and mined only in one part of Derbyshire in any quantity.

BOLT: Metal expansion device fitted into a drilled hole in the rock to make a belay (qv) point.

BONE CAVE: Cave in which archae-ologically significant bones have been found.

BOSS: Stubby stalagmite (qv).

BOULDER CHOKE: Large area of rock collapse partially or completely blocking a cave passage or chamber.

BUDDLE: (mining) Trough or pond in which ore is washed.

CABAN: (mining) Shed, often of stone, in Welsh slate mines, used for meal breaks and meetings.

CALCITE: Crystalline form of calcium carbonate, and the basic constituent of most cave formations (qv).

CARBIDE LAMP: See Acetylene Lamp.

CAVE: Natural cavity in rock – usually used by cavers to denote one large enough to enter; (verb) to explore such cavities.

CAVE DIVER: Caver who specializes in exploring sumps (qv) which necessitate breathing apparatus.

CAVE PEARLS: Spheres of calcite sometimes formed in active cave pools.

CHAMBER: Large opening in a cave or mine.

CHIMNEY: Narrow vertical fissure; (verb) to climb such a fissure using both walls simultaneously.

CHOKE: See Boulder Choke.

CLINTS: Isolated blocks on exposed limestone plateau.

COLUMN: Formation resulting from the joining of a stalactite and stalagmite (qv).

CRAWL: Low passage necessitating hands-and-knees or even flat-out crawling.

CROSS CUT: (mining) Tunnel driven at right angles to a vein (qv).

CURTAIN: Flat, thin hanging formation, often wavy and translucent.

DEADS: Waste rock in a mine (or cave dig (qv)).

DECORATION: See Formation.

DENEHOLE: Vertical shaft cut into sandstone, opening out at the bottom, often of indeterminate age.

DESCENDER: Mechanical device used for abseiling (qv).

DIG: Excavation made on the surface or underground in the attempt to reach a new cave or new passages.

DIP: The angle of tilt of a rock bed (qv).

DRIPSTONE: General name for formations (qv) caused by falling drops of water.

DRY SUIT: Completely waterproof sealed suit.

DUCK: The point at which a cave passage roof comes to within a very short distance of a water surface, passable by cavers (see Free Dive).

EARTH HOUSE: Scottish name for an old, man-made, underground structure, usually for habitation.

ELECTRON LADDER: Modern form of flexible caving ladder with alloy rungs and steel wire sides; can be coiled.

EXPOSURE: Physiological effects of the lowering of the body's critical core temperature by the effects of cold, wet conditions or inadequate food intake.

FACE: (mining) The place in a mine where a vein or seam (qv) is being worked.

FALSE FLOOR: Layer of calcite (qv) left stretching from a cave wall when the underlying supporting material is washed away.

FAULT: Fracture line in rock.

FIRESETTING: (mining) Old mining technique in which the rock is heated by a fire then suddenly quenched with water, producing fracturing.

FLOWSTONE: Calcite covering on wall or floor of a cave.

FOGOU: Old Cornish underground structure, possibly with ritual significance.

FORMATION: Any one of the many forms of cave decoration, chiefly by the deposition of calcite (qv).

FREE DIVE: A cave sump (qv) which can be dived without the need for breathing apparatus.

GALENA: (mining) The common ore of lead.

GINGING: (mining) Dry-stone lining in a mine shaft.

GOUR POOL: Pool created by the formation of a natural calcite dam built up by water flowing over its lip.

GRIKES: The joints surrounding clints (qv).

GROTTO: Small well-decorated chamber in a cave, or a man-made cavity, usually decorated with sea shells.

GYPSUM: Hydrated calcium sulphate, responsible for certain cave formations, and a water soluble rock in which caves can form.

HARNESS: Strapping arrangement worn by cavers for single rope techniques (qv).

HELICTITES: Eccentric cave formations growing apparently in defiance of gravity.

HYPOGEUM: Underground man-made excavation, usually for religious use.

HYPOTHERMIA: See Exposure.

ICE CAVE: Cave formed completely within the ice of a glacier, or one in alpine limestone lined with substantial amounts of ice.

INCLINE: (mining) Sloping passage.

JOINT: Vertical division in rock stratum through to the bedding plane (qv).

KARABINER: Metal snap-link, usually oval, used for fastening ropes, ladders, etc. in caving. (Colloquial: krab.)

KARST: Typical exposed limestone terrain.

KIBBLE: (mining) Ore bucket.

LEVEL: See Adit.

LIFELINE: The rope, controlled by a lifeliner, securing a climbing caver.

LIMESTONE: The rock, at least half of which consists of calcium carbonate, in which most true caves are found.

LODE: See Vein.

MASTER CAVE: The main drainage channel of a cave.

Glossary

MAYPOLE: Metal tube assembled from sections used to raise a ladder to a high-level (usually unexplored) passage.

MEANDER: Exaggerated enlargement of cave passage bend.

MEER: (mining) Old Derbyshire miner's measurement of ground, varying from area to area.

MINERAL: Rock of regular and definite chemical composition.

MOONMILK: Uncommon cave formation consisting of a cauliflower-like mass of soft calcite.

OGOF: Welsh name for a cave.

OPENCAST MINING: (mining) Removal of ore preceded by complete removal of overlying layers.

ORE: Mineral that is worth mining.

OUTCROP: Rock exposed at the surface.

OXBOW: Loop of passage abandoned by the stream.

PHREATIC ZONE: Permanently water-filled zone in a cave and surrounding rock. A phreatic passage is one formed in conditions of complete saturation.

PIPE VEIN: (mining) Natural cavity infilled by minerals.

PITCH: Vertical descent in a cave.

POTHOLE/POT: Cave in which there are vertical shafts; or, such a shaft.

PRUSIKING: Technique used to climb up a fixed rope using mechanical gripping devices or special gripping knots.

RAKE: (mining) Vertical vein of ore.

RESURGENCE/RISING: The point at which an underground stream emerges at the surface.

RIMSTONE POOL: See Gour Pool.

RISE: (mining) Underground shaft excavated upwards.

RUCKLE: Mendip cavers' term for a boulder choke (qv).

SCALLOPS: Oval concave water-flow markings on the walls of a cave passage.

SCRIN: (mining) Small vein of ore, usually parallel with a larger vein.

SEAM: See Vein.

SHAKEHOLE: Depression in a limestone region possibly indicating a cave entrance or the collapse of a cave roof, sometimes taking a stream.

SINGLE ROPE TECHNIQUES (SRT): Those applied in the descent (abseil) or ascent (prusik) of a fixed rope on a cave pitch (qv).

SINK: See Shakehole.

SOUGH: (mining) Drainage tunnel.

SPELEOLOGIST: Generally, in Britain, a caver specializing in one or more of the cave sciences; more loosely, a caver.

SQUEEZE: Tight or awkward constriction in a cave passage.

STALACTITE: Cave formation hanging from the roof of a cave.

STALAGMITE: Cave formation growing up from the floor of a cave, often beneath a stalactite.

STEMPLE: (mining) Usually a wooden beam jammed across a mine passage or shaft, often an aid to climbing.

STOPE: (mining) The hole in a mine from which ore has been, or is being, mined.

STRAW: Thin hollow stalactite (qv).

SUMP: Completely flooded section of cave passage, sometimes passable by diving.

SWALLET: See Shakehole.

TRAVERSE: A horizontal climb at high level in a cave passage or chamber.

TROGLODYTE: Person who lives in a cave.

VADOSE CAVE: The part of a cave formed by a stream with a free airspace above.

VADOSE ZONE: The area lying above the water table.

VEIN: (mining) The part of rock strata that contains ore.

WETSUIT: Protective insulating suit of neoprene foam rubber. Water absorbed by the foam cells is kept warm by body heat.

WINZE: (mining) Underground shaft excavated downwards.

Further reading

Caving: regional guide books

'Northern Caves' series, various authors, Dalesman Books:
Vol 1, *Wharfedale & Nidderdale*, 1979
Vol 2, *Penyghent & Malham*, 1982
Vol 3, *Ingleborough*, 1981
Vol 4A, *Scales Moor & Kingsdale*, 1983
Vol 4B, *Leck & Casterton Fells*, 1983
Vol 5, *The Northern Dales*, 1977
Nicholas Barrington and William Stanton, *Mendip, The Complete Caves and a View of the Hills*, Cheddar Valley Press, 1977
Trevor D. Ford and David W. Gill, *Caves of Derbyshire*, Dalesman Books, 1979
Dave Irwin and Tony Knibbs, *Mendip Underground – A Caver's Guide*, Mendip Publishing, 1985
C. A. Self (compiler), *Caves of County Clare*, University of Bristol Speleological Society
Tim Stratford, *Caves of South Wales*, Cordee, 1978

Mining

A. R. Griffin, *Coalmining*, Longman, 1971
A. R. Griffin, *The British Coalmining Industry*, Moorland, 1977
Geoff Preece, *Coalmining*, Salford Museum of Mining, 1981

Caving

Norbert Casteret, *Ten Years Under the Earth*, Mendip Publishing (Cleeve House, Theale, near Wedmore, Somerset), 1975
Pierre Chevalier, *Subterranean Climbers*, Mendip Publishing, 1975
Martyn Farr, *The Darkness Beckons*, Diadem Books, 1980
Martyn Farr, *The Great Caving Adventure*, Oxford Illustrated Press, 1984
Robert de Joly, *Memoirs of a Speleologist*, Mendip Publishing, 1975
David Judson (editor), *Caving Practice and Equipment*, David and Charles, 1984
J. Lawrence and R. W. Brucker, *The Caves Beyond*, Mendip Publishing, 1975
Jim Lovelock, *A Caving Manual*, Batsford, 1981
Ben Lyon, *Venturing Underground*, EP Publishing, 1983
Mike Meredith, *Vertical Caving*, available in UK from Mendip Publishing.
Descent – the Caver's Magazine, bimonthly, Ambit Publications Ltd, 13–15 Stroud Road, Gloucester, concerns itself particularly with the sport of caving.

Index

Numbers in italics refer to illustrations.

Acknowledgements

The author expresses his thanks to all those who have assisted him with the preparation of this book. His particular gratitude, for their eleventh-hour help, goes to: Dr Paul Cornelius, Ian Davinson, Paul Deakin, John Forder, Clive Gardener, Bill Gascoyne, Dave Gill, Dr Peter Glanvill, Chris Howes, Alan Jeffreys, Bill Maxwell, Dave Mills, Harry Pearman, Paul Tarrant, Clive Westlake, Jerry Wooldridge, Ivan Young, and yet another.

Picture Credits

Dr. A. E. Annels 121; **Big Pit Mining Museum** 111, 112, 114; **Blue John Mine** 131, 132; **British Tourist Association** 116, 117; **J. Allan Cash** 9, 10, 13, 17, 56, 57, 58, 108, 109, 115; **CEGB** 128, 129; **Chatterley Whitfield Mining Museum** 154, 154, 155; **Cheddar Gorge Caves** 81, 82, 84, 85; **Chislehurst Caves** 102, 104, 105; **Clearwell Caves** 92. 94; **Cruachan Hydro-Electric Power Station** 162, 163; **Dan-yr-Ogof Caves** 118, 119; **Ian Davinson** 8, 27; **P. R. Deakin** 18, 35, 39, 41, 44, 46, 51, 52, 78, 144, 145; **Dolaucothi Gold Mine** 122, 123; **Dave Elliot** 140; **John Forder** 45; **George & Charlotte Caves** 74; **Peter Glanvill** 63, 64; **Goodluck Lead Mine** 146, 147; **Hastings Borough Council** 97, 98; **Highlands & Islands Development Board** 164; **Chris Howes** 11, 14, 21, 22, 23, 24, 37, 42, 43, 49, 53, 59, 60, 61, 124, 125; **Ingleborough Caves** 156, 157; **Kents Cavern** 19, 72, 73; **Kitley Caves** 76; **Llechwedd Slate Mine** 126, 127; **Poldark Mine** 69, 70, 71; **Poole's Cavern** 150, 151; **Radio Times Hulton Picture Library** 16, 80; **Royston & District Local History Society** 107; **Scott's Grotto** 99, 100, 101; **Speedwell Cavern** 138, 139; **Stump Cross Caverns** 160, 161; **Welsh Tourist Board** 120; **Clive Westlake** 36, 62; **Wookey Hole Caves** 86, 88, 89, 90, 91; **Jerry Wooldridge** 7, 12, 20, 25, 29, 32, 47, 50, 54, 133, 136, 137, 141, 142, 143, 158, 159; **Ivan Young** 65.